NEW AGE ANALYTICS

Transforming the Internet through
Machine Learning, IoT, and Trust Modeling

NEW AGE ANALYTICS

Transforming the Internet through
Machine Learning, IoT, and Trust Modeling

Edited by
Gulshan Shrivastava, PhD
Sheng-Lung Peng, PhD
Himani Bansal, PhD
Kavita Sharma, PhD
Meenakshi Sharma, PhD

APPLE ACADEMIC PRESS

Apple Academic Press Inc.
4164 Lakeshore Road
Burlington ON L7L 1A4
Canada

Apple Academic Press, Inc.
1265 Goldenrod Circle NE
Palm Bay, Florida 32905
USA

Exclusive worldwide distribution by CRC Press, a member of Taylor & Francis Group

No claim to original U.S. Government works

International Standard Book Number-13: 978-1-77188-875-2 (Hardcover)
International Standard Book Number-13: 978-1-00300-721-0 (eBook)

Library and Archives Canada Cataloguing in Publication

Title: New age analytics : transforming the internet through machine learning, IoT and trust modeling/ edited by Gulshan Shrivastava, PhD, Sheng-Lung Peng, PhD, Himani Bansal, PhD, Kavita Sharma, PhD, Meenakshi Sharma, PhD.
Names: Shrivastava, Gulshan, 1987- editor. | Peng, Sheng-Lung, editor. | Bansal, Himani, 1985- editor. | Sharma, Kavita, 1985- editor | Sharma, Meenakshi, editor.
Description: Includes bibliographical references and index.
Identifiers: Canadiana (print) 20200216945 | Canadiana (ebook) 20200217038 | ISBN 9781771888752 (hardcover) | ISBN 9781003007210 (ebook)
Subjects: LCSH: Machine learning
Classification: LCC Q325.5 .N49 2020 | DDC 006.3/1—dc23

CIP data on file with US Library of Congress

Apple Academic Press also publishes its books in a variety of electronic formats. Some content that appears in print may not be available in electronic format. For information about Apple Academic Press products, visit our website at **www.appleacademicpress.com** and the CRC Press website at **www.crcpress.com**

About the Editors

 **Gulshan Shrivastava**, PhD, is currently working as a research fellow in the department of computer science engineering in the National Institutes of Technology Patna (Institute of National Importance), India. He has edited or authored more than six books, 10 book chapters, and 35 articles in international journals and conferences of high repute, including those published by IEEE, Elsevier, ACM, Springer, Inderscience, IGI Global, Bentham Science etc. He has also served as Associate Editor of several journals, including *the Journal of Global Information Management and International Journal of Digital Crime and Forensics* (published by IGI Global) and Section Editor of SCPE. He is also serving in many various roles for several for journals, such as guest editor, editorial board member, international advisory board member, and reviewer board.

In addition, he received Master of Technology in Information Security from GGSIPU Delhi, India; and Bachelor of Engineering in CSE from the Maharishi Dayanand University, India. He submitted his PhD Thesis in Computer Science and Engineering of National Institute of Technology Patna (Institute of National Importance), India. He also earned SCJP 5.0 certification from Sun Microsystem. He has also delivered expert talks, guest lectures at international conferences and has been a convener for the International Conference On Innovative Computing and Communication (2020 and 2019), organizing chair in the International Conference on Computational Intelligence and Data Analytics (2018), and publication chair for MARC (2018). He is a life member of the International Society for Technology in Education and a professional member of IEEE, CSI, ACM, SIGCOMM, and many other professional bodies. He has an ardent inclination toward the field of analytics and security. His research interests include information security, data analytics, cyber security and forensics, mobile computing, intrusion detection, and computer networks.

Sheng-Lung Peng, PhD, is a Professor in the Department of Computer Science and Information Engineering at National Dong Hwa University, Hualien, Taiwan. He is now the Dean of the Library and Information Services Office of NDHU (NDHU), an honorary Professor of Beijing Information Science and Technology University of China, and a visiting Professor at the Ningxia Institute of Science and Technology of China. He serves the director of the ICPC Contest Council for the Taiwan region, a director of the Institute of Information and Computing Machinery, of the Information Service Association of Chinese Colleges, and of the Taiwan Association of Cloud Computing. He is also a supervisor of the Chinese Information Literacy Association, of the Association of Algorithms and Computation Theory, and of the Interlibrary Cooperation Association in Taiwan. His research interests are in designing and analyzing algorithms for bioinformatics, combinatorics, data mining, and networks. Dr. Peng has edited several special issues of journals, such as *Soft Computing, Journal of Internet Technology, Journal of Computers, MDPI Algorithms,* and others. He published over in 100 journal and international conferences papers.

Dr. Peng received a BS degree in Mathematics from National Tsing Hua University, and MS and PhD degrees in Computer Science from the National Chung Cheng University and National Tsing Hua University, Taiwan, respectively.

Himani Bansal, PhD, has served as an Assistant Professor at various reputable engineering colleges and universities. Currently, she is working with the Jaypee Institute of Information Technology, Noida, India. She has over 12 years of demonstrated success and an excellent academic record (batch topper in master's and 100/100 in Mathematics in Class X). Her extensive background of includes involvement in team building activities, effective communication with excellent planning, organizational, and negotiation strengths, as well as the ability to lead, reach consensus, establish goals, and attain results. Her many reputed certifications include SAP-ERP

Professional from the prestigious company SAP India Pvt. Ltd.; UGC National Eligibility Test (NET), IBM Certified Academic Associate DB2 9 Database and Application Fundamentals; Google Analytics Platform Principles by Google Analytics Academy; E-Commerce Analytics by Google Analytics Academy; and RSA (Rational Seed Academy) by IBM India. She has penned many research papers for various international journals and conferences. She also has books to her credit along with edited chapters in international books. She was awarded the Eighth Annual Excellence Award in Research Journal Awards for the outstanding article by IGI Global for "Mitigating Information Trust-Taking the Edge of Health Websites." She has served as a section editor, guest editor, convener, and session chair for various reputable journals and conferences, such as SCPE, NGCT, India Com, CSI Digital Life, IJAIP, JGIM, ICACCI, ICCCA, etc. and has reviewed many research papers. She has organized, coordinated, and attended FDPs, trainings, seminars, and workshops. She has also coordinated workshops for IIT Bombay under the NMEICT program initiated by the Ministry of HRD, India. She serves as a life member of various professional societies, such as CSI, ISTE, CSTA, and IAENG, and is an active member of IEEE and ACM.

Dr. Bansal earned her engineering degree from Rajasthan University in 2006, followed by corporate experience with a reputed company. Further, to fulfill her academic interests, she completed her master's and PhD degrees from Birla Institute of Technology, Mesra, India.

Kavita Sharma, PhD, is working as an Associate Professor in the Department of CSE at G.L. Bajaj Institute of Technology & Management, Greater Noida, India. She formerly worked as an Assistant Professor at Dronacharya College of Engineering, Greater Noida, India, and Lecturer in Northern India Engineering College, Delhi, India. She has published five books and more than 45 research papers, chapters, editorials in international journals and conferences. She is also served as Section Editor of *Scalable Computing* (SCPE). She is also serving many reputable journals as guest editor, editorial board member, and international advisory board member. In addition, Dr. Sharma has also delivered the expert talks and guest lectures at international conference and serving as the reviewer for journals of IEEE, Springer, Inderscience, Wiley, etc. She is a member of IEEE, ACM, CSI (Life Member), SDIWC, Internet Society, Institute of Nanotechnology (Life Member), IAENG (Life Member), IACSIT, CSTA, and IAOE. She has actively participated in different faculty development programs and in various national and international workshops. Her areas of interest includes information and cyber security, mobile computing, android, web mining, data analytics and machine learning.

Dr. Sharma received a PhD in Computer Engineering from the National Institute of Technology, Kurukshetra, (Institution of National Importance) India, and an MTech in Information Security from Ambedkar Institute Technology of Advanced Communication Technology & Research (formerly Ambedkar Institute of Technology), Delhi, India. She has also completed her BTech in IT from the I.M.S. Engineering College, Ghaziabad, India. In addition, she was awarded a Fellowship from the Ministry of Electronics and Information Technology, Government of India.

Meenakshi Sharma, PhD, is working as an Professor at the School of Computer Science and Engineering, Galgotias University, India. She has over 15+ years of experience in teaching. She has published over 50 research publications in *IEEE Transaction, SCIE, SCI,* and Scopus. She worked as a guest editor for the *International Journal of Electronics, Communications, and Measurement Engineering.* She had guided one PhD candidate, and six candidates are currently under her guidance. Her research interests are big data analytics, data compression, digital image processing, and data warehousing. She received her MTech in Computer Science and Engineering from Kurukshetra University, India, and was awarded her PhD in Computer Science from Kurukshetra University as well.

Contents

Contributors

Ravin Kaur Anand
Division of Information Technology, Netaji Subhas Institute of Technology, New Delhi, India,
E-mail: ravinkaur97@gmail.com

Rohit Anand
Department of ECE, G. B. Pant Engineering College, New Delhi, India,
E-mail: roh_anand@rediffmail.com

Himani Bansal
Department of CSE, Jaypee Institute of Information Technology, Noida, India,
E-mail: singal.himani@gmail.com

Ruchika Bathla
Amity Institute of Information Technology, Amity University, Noida, Uttar Pradesh, India,
E-mail: bathla.ruchika@gmail.com

Nidhika Chauhan
University Institute of Computing, Chandigarh University, India,
E-mail: nidhi29.chauhan@gmail.com

Nikita Chawla
Division of Information Technology, Netaji Subhas Institute of Technology, New Delhi, India,
E-mail: nikitachawla3@gmail.com

Prasenjit Choudhury
Department of Computer Science and Engineering, National Institute of Technology Durgapur, India

B. D. Deebak
School of Computer Science and Engineering, Vellore Institute of Technology, Vellore–632007,
India, E-mail: deebak.bd@vit.ac.in

Anshul Garg
Department of Computer Applications, Chandigarh Group of Colleges, Landran (Mohali),
Punjab, India, E-mail: mca.anshulgarg@gmail.com

B. Gomathy
Department of Computer Science and Engineering, Bannari Amman Institute of Technology,
Tamil Nadu, India, E-mail: bgomramesh@gmail.com

Ankita Gupta
Department of ECE, G. B. Pant Engineering College, New Delhi, India

Amit Jain
Associate Professor, University Institute of Computing, Chandigarh University, Chandigarh, India,
E-mail: amit_jainci@yahoo.com

Nitin Jain
School of Computer Science and Engineering, Galgotias University, Greater Noida, UP, India

Abhishek Majumder
Department of Computer Science and Engineering, Tripura University (A Central University),
Tripura, India, E-mail: abhi2012@gmail.com

Pragi Malhotra
Division of Information Technology, Netaji Subhas University of Technology,
(Formerly Netaji Subhas Institute of Technology), New Delhi, India,
E-mail: pragi18@gmail.com

Deepa Mehta
Department of CSE, SEST Jamia Hamdard, India, E-mail: deepa.mehta12@gmail.com

S. Muthuramalingam
Department of Information Technology, Thiagarajar College of Engineering, Madurai,
Tamil Nadu, India, E-mail: smrit@tce.edu

Anand Nayyar
Graduate School, Duy Tan University, Vietnam, E-mail: anandnayyar@duytan.edu.vn

O. Obulesu
Department of Computer Science and Engineering, Malla Reddy Engineering College
(Autonomous), Secunderabad, India, E-mail: oobulesu681@gmail.com

Chhabi Rani Panigrahi
Department of Computer Science, Rama Devi Women's University, Bhubaneswar, India,
E-mail: panigrahichhabi@gmail.com

Bibudhendu Pati
Department of Computer Science, Rama Devi Women's University, Bhubaneswar, India,
E-mail: patibibudhendu@gmail.com

Sheng-Lung Peng
Department of Computer Science and Information Engineering, National Dong Hwa University,
Hualien, Taiwan, E-mail: slpeng.@ndhu.edu.tw

Pijush Kanti Dutta Pramanik
Department of Computer Science and Engineering, National Institute of Technology Durgapur,
India, E-mail: pijushjld@yahoo.co.in

Ritu Punhani
Department of Information Technology, ASET, Amity University, Noida, Uttar Pradesh, India,
E-mail: ritupunhani@gmail.com

S. Rakeshkumar
Department of Computer Science and Engineering, GGR College of Engineering, Vellore,
Tamil Nadu, India, E-mail: rakesherme@gmail.com

V. Ramasamy
Department of Computer Science and Engineering, Park College of Engineering and Technology,
Tamil Nadu, India, E-mail: researchrams@gmail.com

Sonia Saini
Amity Institute of Information Technology, Amity University, Noida, Uttar Pradesh, India,
E-mail: sonia.22.saini@gmail.com

Joy Lal Sarkar
Department of Computer Science and Engineering, Tripura University (A Central University),
Tripura, India, E-mail: joylalsarkar@gmail.com

Deepak Kumar Sharma
Division of Information Technology, Netaji Subhas University of Technology,
(Formerly Netaji Subhas Institute of Technology), New Delhi, India,
E-mail: dk.sharma1982@yahoo.com

Kavita Sharma
National Institute of Technology, Kurukshetra, India
E-mail: kavitasharma_06@yahoo.co.in

Meenakshi Sharma
School of Computer Science & Engineering, Galgotias University, India,
E-mail: minnyk@gmail.com

Gulshan Shrivastava
Department of Computer Science & Engineering, National Institute of Technology Patna, India
E-mail: gulshanstv@gmail.com

Pradeep Kumar Singh
Department of Computer Science and Engineering, National Institute of Technology Durgapur, India

Ankit Srivastava
Department of ECE, G. B. Pant Engineering College, New Delhi, India

Loredana Stanciu
Politehnica University Timisoara, 2 Victory Square, 300006 Timisoara, Romania,
E-mail: loredana.stanciu@aut.upt.ro

Tina Tomažič
Institute of Media Communications, University of Maribor, Slovenia,
E-mail: tina.tomazic@um.si

Jayant Verma
Division of Information Technology, Netaji Subhas University of Technology,
(Formerly Netaji Subhas Institute of Technology), New Delhi, India,
E-mail: jayantverma1998@gmail.com

Sherin Zafar
Department of CSE, SEST Jamia Hamdard, India, E-mail: zafarsherin@gmail.com

Abbreviations

ACS	adjusted cosine similarity
AI	artificial intelligence
AIC	Akaike Information Criteria
ANNs	artificial neural networks
AUA	American Urological Association
AUC	area under curve
B2B	business to business
B2C	business to consumer
B2G	business to government
BA	business analytics
BD	big data
BGFS	backward greedy feature selection
BIC	business, innovation, and customer experience
BP	business processes
BPMN	business process modeling notations
C2B	consumer to business
C2C	consumer to consumer
CA	cellular automata
CA	context-aware
CART	classification and regression trees
CB	content-based
CC	clustering coefficient
CDC	Center for Disease Control and Prevention
CF	collaborative filtering
CFW	correlation-based feature weighting
CS	cosine similarity
CVP	customer value proposition
DAG	directed acyclic graph
DoS	denial-of-service
DT	decision tree
ED	Euclidean distance
ESA	explicit semantic analysis
ESMOS	epidemic sentiment monitoring system

FCC	Federal Communication Commission
FCV	fold cross-validation
FOAF	friend-of-a-friend
GPUs	graphics processing units
GSI	geological survey of India
HB	hybrid filtering
HIPAA	health insurance portability and accountability act
HMMs	hidden Markov models
IDE	infrastructure-driven epicenter
IEC	International Electro-Technical Commission
IEEE	Institute of Electrical and Electronics Engineer
IETF	internet engineering task force
IG	information gain
IMPACT	Information Marketplace for Policy and Analysis of Cyber-Risk and Trust
IoT	internet of things
IPFIX	IP flow information export
ISBA	item similarity-based approach
ISO	International Standard Organization
ISPs	information service providers
ISTS	implicit social trust and sentiment
IT	information technology
ITU	International Telecommunication Union
JDBC	java database connectivity
JS	Jaccard similarity
KB	knowledge-based
KDD	knowledge discovery in databases
KNN	K-nearest neighbor
LR	logistic regression
LT	long tail
MAE	mean absolute error
MAP	maximum A posteriori
MC	mean centering
MD	Manhattan distance
ML	machine learning
MLP	multi-layer perceptron
MLR	multinomial logistic regression
MLWSVM	multilevel weighted SVM

MOP	multiple open platform
MQTT	message queuing telemetry transport
MSD	mean squared distance
NB	Naive Bayes
NFC	near field communication
NLP	natural language processing
NN	neural network
OCkNN	outsourced collaborative kNN protocol
OEMs	original equipment manufacturers
OLS	ordinary least square
OM	opinion mining
OSNs	online social networks
PaaS	product-as-a-service
PARDES	publish-subscribe applied to distributed resource scheduling
PB	PARDES broker
PC	Pearson correlation
PCA	principal component analysis
PIDD	Pima Indian diabetes data set
POS	point of sale
PS	percent-age split
QoS	quality of services
REST	representational state transfer
RF	random forest
RFCs	requests for comments
RFID	radio frequency identification
RL	reinforcement learning
RMSE	root mean square error
ROA	resource oriented approach
ROC	receiver operating characteristic
RS	recommender system
SA	sentiment analysis
SC	spearman correlation
SCM	supply chain management
SDK	software development kits
SIoT	social internet of things
SNMP	simple network management protocol
SNPs	single nucleotide polymorphism
STB	set-top box

SVM	support vector machine
SWRL	semantic web rule language
TAG	traffic activity graphs
TAM	trust association and management
TAs	trustworthiness attributes
TCG	trusted computing group
TCP	tobacco control program
TE	technology enabler
TIs	trust indicators
TPM	trusted platform module
TPUs	tensor processing units
TSF	time series forecasting
URIs	uniform resource identifier
USBA	user similarity-based approach
UTD	use training data set
VMs	virtual machines
WA	weighted average
WoT	web of things
WS	web services
WSDL	web services description language
WSNs	wireless sensor networks
WSoT	web-stack of things
ZS	Z score

Preface

"A baby learns to crawl, walk, and then run. We are in the crawling stage when it comes to applying machine learning."

—Dave Waters

This book will not only help you to crawl, but also can lead you to the first few steps of the baby walk. It deals with the importance of tools and techniques used in machine learning. Each and every chapter has important aspects in various application areas in machine learning. The book explains how advancement in the world of the web has been achieved and the experiences of the users have been analyzed. Initiating from data gathering of all volumes by all electronic means to story-telling by the data, every aspect has been discussed. How whole data is analyzed and managed, voluminous data concealed, erected, and nodded explaining the response of user and everything is talked about. Connectivity through hyperlinks for safe shopping is one of the key areas touched. Modeling for online shopping is also discussed. The Concept of Quality of Service (QoS) for internet safety points to the limits and value of information sharing. Recommender systems with collaborative filtration and IoT framework along with various payback models are discussed with the help of case studies. The plan adopted to enumerate the frequency to evaluate data according to the result is also verified. The relation between the problem-solving business and the purchaser is also highlighted. The importance of the data sets that gives information and business analysis using big data is also present in the book. Full references of sources (original) have been discussed and verified.

Chapter 1 by Sharma et al. presents information of cyber-sphere that has paved the way for the global transformation. It explains the provocations brought by the internet with the introduction of the scrutiny of gathered data. The devices and proficiencies of information conclave are rigorously inspected and explained.

Chapter 2 by Ramasamy et al. notes the numerous categories of the algorithms of the machine learning. They explain the tools and techniques associated with machine learning and where data gathered in an electronic way forecasts the time ahead.

Chapter 3 by Jain et al. explains the logic at the pretext of turning the huge bulk of information accessible for alteration in proficiency with the evolution of diverse advancements in the approaches, which helps to meet the provocation in the coming times. This chapter is about the sheer magnitude of analyzing the different facets in the context of mechanized effects in the management of networking.

Chapter 4 by Sharma et al. describes the principles of data mining in the times when data is getting bigger and bigger. It gives the sound understanding of finding various concealed logic erecting from the hyperlinks in the nodes of networking that reposes the coupled data that is needed at the user end.

Chapter 5 by Sharma et al. discusses the rudimentary principles of social pursuit and its connectivity with humans. It also explains the safe online domain, considering the risk management having various options for online shopping.

Chapter 6 by Kumar et al. emphasizes the conviction of users in online shopping, which entangles networking with social sites, including the sentiment modeling approach. Working on the procedure and its dependencies is explored to a large extent.

Chapter 7 by Zefer et al. is based on Quality of Service (QoS) and explains the various attacks on cyber, enumerating the ways to enhance trust in the cyber world. The authors have highlighted some elementary points that inspect the hubs and set a limit value to share the information with different groups and the different gadgets.

Chapter 8 by Singh et al. deals with the consequences of the recommender system, which is a technique based on collaborative filtering. New ways for e-commerce that uses RS as its base for online services are explained. It helps the customers to find a quality product to buy online by checking its ranking in terms of its ratings using various research methodologies.

Chapter 9 by Sharma et al. discusses the different frameworks of business models of IoT with the help of a case study. It entails the practical conception that how IoT works. It also includes the terminologies used with various applications that can be used for various aspects such as e-commerce and m-commerce.

Chapter 10 by Gupta et al. defines the importance of a set of accurate data that explains all the features, so that a good decision can be made by the user. It helps in the advancement of business by using the information stored. A case study is discussed dealing with big data, IoT, and Business Analysis (BA), which is the major highlight of this chapter.

Chapter 11 by Deepak et al. discusses the relation of a purchaser and gives the lumps of business with problem-solving. It signifies the uncoupled aims. Different models have been used to clarify the topic. A comparative study is given for the inscription of all the business models.

Chapter 12 by Saini et al. has considered the numerous issues related to the cyber world in consideration of health issues. Projections have been used to enumerate the frequencies. The data has been taken through social media to get the specified result.

—**Gulshan Shrivastava**
National Institute of Technology Patna, India

—**Sheng-Lung Peng**
National Dong Hwa University, Hualien, Taiwan

—**Himani Bansal**
Jaypee Institute of Information Technology, Noida, India

—**Kavita Sharma**
G. L. Bajaj Institute of Technology & Management, Greater Noida, India

—**Meenakshi Sharma**
Galgotias University, India

CHAPTER 1

Digital Marketing and Analysis Techniques: Transforming Internet Usage

MEENAKSHI SHARMA,[1] NIDHIKA CHAUHAN,[2] HIMANI BANSAL,[3] and LOREDANA STANCIU[4]

[1]Galgotias University, India, E-mail: minnyk@gmail.com (M. Sharma)

[2]Chandigarh University, Chandigarh, India

[3]Jaypee Institute of Information Technology, Noida, India

[4]Politehnica University, Timisoara, Romania

ABSTRACT

In the era of immense digitalization, changes have occurred in the field of analysis. Data analysis has boost up opportunities not only in business but in another field as well. With the introduction of the internet, the world has undergone various transformations. This chapter mainly focuses on how new-age transformation is changing the Internet. It also reflects the opportunities for analysis and showcases the changes brought to our everyday life by the internet. As it is known the internet is a massive source of data gathering, this gathered data is given some form by analysis. Data analysis has become a popular term these days and has opened new career opportunities. Various tools and techniques of data gathering and analysis will be discussed in this chapter.

1.1 INTRODUCTION

Internet and data analysis are two interlinked terms but this does not mean that data analysis is only restricted to data gathering and works only on existing data, but it is also about prediction and forecasting. The best example is the stock exchange forecast or weather forecast. In order to make the prediction both fields use analysis. The analysis is not only restricted to these fields, but there are many more areas where analysis is used, like in business it plays a considerable role and helps in boosting the business.

Now the word analysis may sound very easy. People might think that all required to do is pick the data from the internet, feed it to the system and them glance at the results and make the prediction. Though the steps are the same, there are many steps in between which involve a collection of data, cleaning of the data, slicing, and dicing of data and much more which makes this process tedious job but thanks to new-age tools and software that have given some relaxation. A details description will be given later in the chapter that will be discussing the challenges with data collected over the internet. Enormous challenges were faced in the case of big data (BD), which will be explained in detail later.

In order to understand the concept, there needs to be clarity of basic terms like analysis, how the internet is linked to an analysis, BD, structured, and unstructured data, various analysis tools, etc. With the exploration of Hadoop, the information system has taken a different turn and information architecture has undergone a considerable transformation (Robert, 2015). For ages, business analysis and intelligence were dependent on the organization's data mart and data warehouse. Some best data analysis was also defined for them; however, there has been a noticeable change in this system. Now the area of analysis and data storing is transforming. With the immense growth of data, the techniques to handle the data and to analysis are also transforming. The internet has been an enormous source of data for us, but the problem lies in structuring this data as it is not homogeneous in nature. Now, the question is why to require the homogeneous data. So in order to perform the analysis, the data is to be clustered in groups which means the data has to be arranged in homogeneous sets. Similar kinds of data when putting together results in easy analysis.

The semantic web is an environment which helps human and machine to communicate semantically. It mainly focuses on making the content suitable for a machine to understand it, so as to extract query (Sharma et al., 2012)

Although there are numerous advantages to this digital transformation, there are many obstacles to this transformation that cannot be ignored. To understand these obstacles there needs to understand the fundamental transformation, how the system was evolved and now where have humans reached and what are the other challenges faced today. A diagrammatic representation of the initiating digital transformation journey gives a clear picture of how an organization responds to such changes (Figure 1.1).

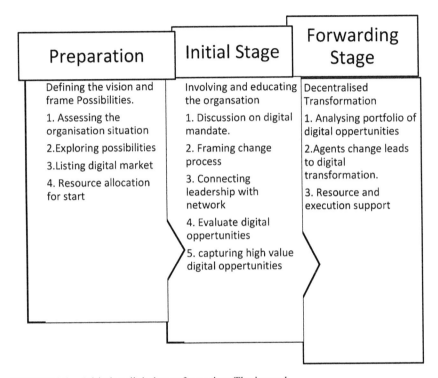

FIGURE 1.1   Initiating digital transformation: The journal.

The digital transformation was not an overnight process; a lot of planning and structuring was done to move from manual phase to the digital transformation phase. As Figure 1.1 represents transformation involved active participation from the beginning till the end. In the beginning stage, some parameters of digital innovation and changes were set (Figure 1.2).

As it is evident from Figure 1.2 that over the past years the organizations have undergone a considerable transformation at different levels. Apart from traditional methods, organizations have moved to the Internet, CRM, and ERP system which has further led to three different types of experiences. These experiences are basically how the members linked to the organization respond to the transformation (Hinchcliffe, 2018). Internet is a wide source of information that has increased the use of web usage mining techniques in academics and commercial areas. Studies show that intrigued web users give profitable inputs to web designers for efficient organizing of website (Sharma et al., 2011) (Figure 1.3).

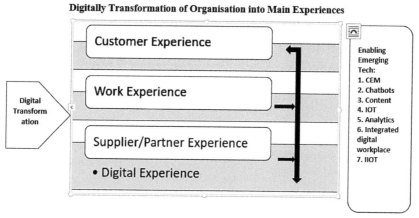

Digitally Transformation of Organisation into Main Experiences

FIGURE 1.2   Digital transformation.

Now moving on to the next phase of transformation framework. Figure 1.3 clearly shows the generative factors which means the source that provides the input. Now having such a system is itself a critical task and getting people to function across this framework is another tedious job. The most challenging task is to get organization collaboration across various parameters and without this collaboration, digital transformation is difficult to achieve. It is believed that people will and cannot adapt to digital transformation overnight, so in order to change the mindset of these people, specific activities are proposed like MOOCs, reverse mentoring, certification, etc. thus leading to the building of required digital skills. Once the organization agrees to adopt the digital system, a framework is designed to identify the starting point and then the ongoing steps. Now the question is how the internet is playing a role in this? So in that case, the internet has been a considerable source for data collection and data gathering. Cloud, Hadoop, etc. are used to handle this immense data. The digital transformation model discussed above in Figure 1.3 mainly focuses on day to day functionality and timely alterations are also made to it so as to stay updated and get proper outcomes (Hinchcliffe, 2018).

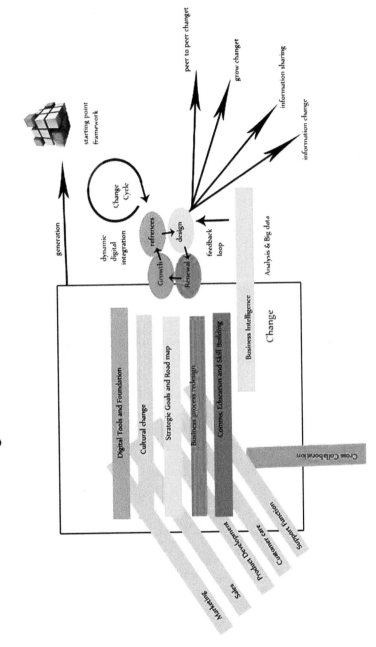

FIGURE 1.3 An adaptable digital transformation framework.

1.2 BUSINESS APPLICATIONS IN SOCIAL MEDIA AND ITS ANALYSIS

Social media is the way forward for Businesses. One of the most prominent uses of social media in businesses is Digital Marketing.

Digital Marketing is a new way for corporations to connect with their audience. It is the sphere that provides equal footing to each business, regardless of its size or resources.

When digital marketing is discussed, then there are many factors that influence it on the ground level to top-level working. They can include factors such as brand communication which can further be used to indicate the traces of digital marketing. However, for the complete evaluation of digital marketing, the explanation should revolve around the working of digital marketing and also its scope in various industries (Fan et al., 2014).

1.2.1 CURRENT DIGITAL MARKETING LANDSCAPE

As the era is emerging, so is the scenario of digital marketing getting its priority. Organizations must observe that the most obvious way to deal with marketing, i.e., via advertising is still ruling over the market with TV and Ads. The available online platforms still need more importance to be given as the industries have realized that online sources have been commonly utilized by the consumers and can be easily targeted for marketing. This automatically will generate more revenue.

There have been five forms of digital marketing.

- Static images take less time to create and are easy to share. The most common content of static images is in the form of infographics or memes.
- Videos or animations have a greater tendency to attract more consumers as people are more attracted to motion. Even after its high efficiency, brands avoid using video content due to its high cost of production.
- The consumers want an easy text to read which gives a minimum appeal.
- Mobile Apps are also considered one of the platforms for content where all content types are provided. It also helps in conveying brand communication (Ruhi, 2014).

- Service provided by the agencies is divided into specific categories. Some of them might provide 360 solutions whereas few charges on a particular specification of a service.

When it comes to services provided by content is the centralized source for communication. Thus, agencies spend more time in this sector. Whereas media helps an organization in structuring their online content.

1.2.2 DIGITAL MEDIA ANALYTICS

When discussing digital campaign then the most important part to notice is Analytics. It has been observed until now that the brands did not hire good analysts. It has started taking shape, as the companies are now aware that they should evaluate these campaigns to run for a marathon. The inbuilt support on Google Analytics to use popular websites named socialbaker. com as the companies have evaluated the impact of their online campaigns (Järvinen et al., 2015).

1.2.2.1 TABLE ACTION SPECIFICATION: PEAK TRAFFIC FOR ONLINE POSTS

The brand has become aware of the best time to launch a product. Different target groups have a different schedule of coming online. One such illustration is mentioned in Figure 1.4. It describes spending hours of consumers online at different time slots.

1.2.2.2 GROWTH OF DIGITAL MOBILE ADVERTISING

The advertising with the help of videos is a new addition in the digital marketing world. It also increases up to the use of different sites primarily in the field of the 3-news portal on Facebook, YouTube, etc.

This can be understood as the viewer will use the portal and without ignoring it, they will watch it. As they will lookout for a way to ignore it, these videos will pop up with a different kind of message with a minimal amount of duration that can vary around 10–20 seconds. The minimum time will undoubtedly lead to a maximized effect that will pop up when the viewer has no clue of it (Smith, 2018). Annual growth of digital ads through mobile in comparison to all other digital sources has been observed to be more in the former which covers a wide range.

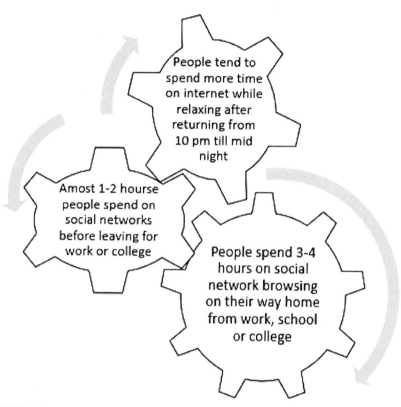

FIGURE 1.4   Illustration of traffic hours.

1.2.3   ORGANIZING SOCIAL MEDIA CONTEST

Brands are more involved in content ideas. The most common and popular contest is the photo or selfie to attract large audience response. An individual has to post their photograph or any other content #tag the brand. The brand not only promotes itself but also generates quality, original content, effortlessly, and without spending many resources (Truong et al., 2010). Different contests are organized to engage more audience on the social media pages which affiliate to different brands. For smart content statistics and computation has been proposed so that the author is identified from his writing style thus ensuring genuine writings (Banga et al., 2018).

1.2.4 E-COMMERCE GROWTH

E-Commerce growth is more of a recent trend, especially with companies such as Alibaba and Amazon pushing to increase their demographic reach. If looked back about 5 to 10 years, e-commerce companies were very distant from being fashionable and limited to only a few developed countries. The reason behind this drastic change is the technical support and increasing credibility of e-commerce. Not to mention, comfortable, and better access to the Internet, has provided a new prospect to the entire outlook of e-commerce. A perfect delivery system is one of the reasons why people are relying more on e-commerce (Bakos, 2001).

1.2.5 NAIVE ADVERTISING TAKING THE MARKET

Advertising can be segregated into many components and one such component is Naive Marketing. Its functions have been cooped up as same as that of advertising. It has further been divided into various forms where the Navie Search Engine is the primary component. Whereas it also has other forms such as Naive News Feed Ads, Naive Advertorial Ads, etc. In the modern era, the rate of native advertising is growing rapidly. The key principle at play here is the collection of data over various platforms, allowing the companies to target the audience based on their previous usage of the internet. In other words, analyzing and evaluation of your information, over time, to tailor the services to yield better results and higher customer satisfaction (Agrawal et al., 2013).

1.2.6 INCREASE IN SMARTPHONE PENETRATION

The mobile market has made a prediction from records that it will raise from 61.1% to 69.4%. This has shown that people are deprived of technology, which is curious to try out new social media platforms in rural areas. Their curiosity motivates them to post photos, connect with friends and family. The excitement of keeping up with what the people around them are doing, this will drive more users on the social platform via mobile (Karim et al., 2013).

1.2.7 PREFERENCE TOWARDS SOCIAL MEDIA PAGES OVER WEBSITES

There are approximately 1.2 million Indian Facebook users who search mainly for the restaurant, business or brand to get the overview and to gain more knowledge as social media is the most preferred which provides the conversational platform. Here the customer can read the comments and can assess the company's image.

A website being is a one-way street demands continuous investments from brands. Social media pages for engagement. Social media has provided the privilege for real-time conversation and this instant responsive system has turned out to be a massive opportunity for companies and also helps in addressing consumer's needs instantaneously within low cost and faster methods (Persaud et al., 2012)

Today social media helps various businesses to connect with the desired audience directly, quickly, and effectively.

1.2.8 EMERGENCE IN LOCAL MARKET

Today social media has grown a lot, there is enormous emergence of various platforms like Twitter which can be used as the medium for announcements on various product releases and another tweet that consumers follow throughout the day. Twitter is not the only platform; Instagram is also one of them which are mainly used for promoting through "photo contents." New businesses can use them for product promotion etc. Next in the line is Pinterest; it is a resource-sharing board. These days Twitter and Instagram are the platforms that are bombarded with a tremendous amount of posts every day and this is the reason why some important posts do not get the attention they deserve. So Pinterest is used to pin resources or information that the client wishes to share with his desired audience without losing them. Fashion or lifestyle boards can be created for various brands and then the consumers follow as per their interest.

Such platforms help in the development of small and localized businesses. Although social media gives an equal opportunity to all, the resources flooded in by the relatively bigger companies can be overwhelming for such businesses that are losing themselves in the crowd.

However, various social media platforms have different mechanisms and algorithms that provide an equal opportunity for such businesses, such as Pinterest (Watson et al., 2013).

1.3 WEB NETWORKING ANALYSIS TOOLS AND SERVICES

Web analysis is the process of data gathering, data examining, data processing, and formulating online strategies for advanced web users. It not only helps in estimating web traffic but also it can be utilized as an apparatus for business and statistical surveying, and to evaluate and enhance the viability of a site in Figure 1.5. Web analysis applications can likewise help organizations measure the consequences of customary print or communicate promoting efforts (Bansal et al., 2018). It encourages one to assess how traffic to site changes after the dispatch of another marketing effort. Web analysis provides data about the number of guests to a site and the number of online visits. It enables the measurement of traffic and recurring patterns that is valuable for any statistical surveying (Watson et al., 2013).

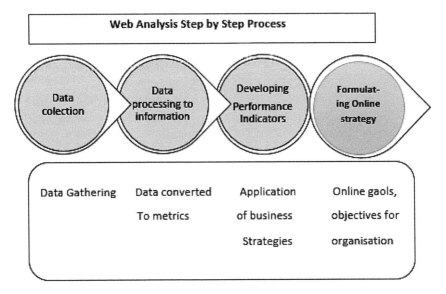

FIGURE 1.5   Web analysis process.

1.3.1   *AI TO INCREASE KNOWLEDGE ON WEB NETWORK ANALYSIS*

Data analysis is a conventional technique for fraud identification. All they need is an analysis technique to study various areas of information like the financial system, business practices, aspects related to management, and legal aspects.

Earlier data analysis strategies were such that focused on fetching factual and quantitative information.

These systems encourage qualitative data elucidations to improve bits of information for further procedures (Gupta et al., 2018).

To go past, a data analysis framework must be furnished with a significant measure of the foundation of information and can perform reasoning assignments including that information and the data provided. In the process, to meet this objective, analysts have diverted to the machine learning (ML) field. This was the only logical opinions since the AI undertaking can be depicted as factories, turning data into useful information. Data or patterns that are novel, legitimate, and conceivably helpful are not merely data, but information.

1.3.2   *TRANSFORMATIVE TECHNIQUES FOR LEARNING THE BEHAVIOR PATTERN FOR BUSINESS APPLICATIONS FROM THE INTERNET*

In artificial intelligence (AI), an evolutionary algorithm is a subset of evolutionary computation; a conventional populace based met heuristic streamlining calculation. An EA utilizes instruments motivated by organic development, for example, generation, transformation, recombination, and choice. Hopeful answers for the streamlining issue assume the job of people in a populace, and the wellness work decides the nature of the arrangements. The development of the populace at that point happens after the rehashed use of the above administrators (Konstas et al., 2009).

1.3.3   *DATA MINING ALGORITHMS FOR SOCIAL NETWORKS USED IN DECISION SUPPORT IN E-BUSINESS APPLICATIONS*

Mining of data is the process just before outcome designs in an extensive pattern of informational collections which includes techniques at the crossroads of AI, database systems, and statistics. Data mining is a subfield of software engineering which separates the raw data from collection processed information transforming the data into the comprehensive structure for further use. Data mining is an embarking investigation e of the "knowledge discovery in databases" (KDD). Apart from critical data analysis, it also includes information with board angles, post-handling of found structures, online updating, perception, information pre-preparing, intriguing quality measurements, model, and deduction contemplations and multifaceted nature contemplations. The distinction between data analysis and data mining is that information examination is to abridge the history, for example, dissecting the viability of an advertising effort, interestingly, data mining centers around utilizing explicit ML and factual models to anticipate the future and find the patterns among information (Yadav et al., 2011).

Figure 1.6 is the representation of the basic data mining process. There are various stages like resource domain which is the primary source for data then all this data is segmented into various databases after which various approaches are used so as to create homogeneous data clutters. The final phase is the method of analysis and the combination of methods results in a summarized report.

1.3.4   MINING AND ANALYZING APPLICATIONS WOT DATA FOR DECISION SUPPORT AND OPTIMIZATION

Present-day businesses utilize a few sorts of facilitator frameworks to encourage information disclosure and support decision making. Value-based application frameworks more often have advanced reports introducing information by utilizing ideas like arranging, gathering, and information accumulation. OLAP frameworks, likewise referred to as an information system, utilize a data warehouse as an information source, and project a superior level tool facilitating in decision making. In such a data warehouse, information is intermittently separated in a collected structure from value-based data frameworks and other outer sources by data warehouse tools. Both, transactional information frameworks and OLAP frameworks, are commonly founded on ideas of arranging, gathering, and information total, where with information collection one of the accumulating capacities like entirety, least, most extreme, tally, and normal are utilized. Both, transactional application frameworks report and OLAP frameworks empower the introduction of various perspectives on collected information in various measurements, the last mentions anyway showing a greater number of measurements than the previous (Provost et al., 2013).

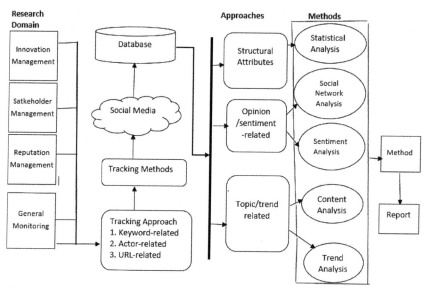

FIGURE 1.6   Data mining process.

1.4   OPTIMIZATION OF DYNAMIC PROCESSES IN WEB FOR BIG DATA (BD) APPLICATIONS

BD includes the gathering, union, and examination of divergent information from different sources and arrangements to uncover new bits of knowledge and convert it to be usable. While BD got its beginning in the purchaser and monetary administration segments, its potential for uncovering profitable data has all the more, as of late caught the consideration of the worldwide manufacturing industry. Without a doubt, BD speaks to a noteworthy segment of Industry 4.0 and the industrial internet of things (IoT). It even gains on the cloud.

Before diving into the manners by which BD can be utilized in procedure improvement, it's profitable to think about why BD is just presently making its passage into the manufacturing domain? While analyzing and answering, the three key variables emerge information, stockpiling, and examination. Figure 1.7 explains the course of recent years, in which the manufacturing business has seen an emotional drop in the expenses of both sensor advances and information stockpiling. Those two improvements have made it financially efficient for producers to gather and store the bulk information that was before achievable. The third factor critical advances in analytical technology, is similarly vital, as creative analyses are driving new prescient analytics and improvement capacities (Luo et al., 2015).

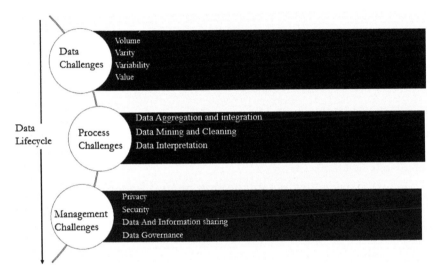

FIGURE 1.7   Data lifecycle.

1.5 IMPLICATIONS OF ELECTRONIC GOVERNMENT

In a study, it was found that the presence of town Internet facilities, offered by the government to residence services was decidedly connected with the rate at which the locals acquire a portion of these services. In a study of a provincial Internet venture, it recognized a positive connection for two such Internet services: acquiring birth endorsements for kids and applications for seniority annuities. Both these taxpayer-supported organizations were of significant social and financial incentive to the nationals. Townspeople report that the Internet-based services spared them time, cash, and exertion and acquiring straightforwardly from the administration office. Therefore, finding states that these services can decrease corruption in the conveyance of this administrative works/service. After more than one year of fruitful results, the e-government program was not ready to keep up the important dimension of nearby political and managerial help to remain institutionally feasible. As government officers moved from the area or developed to discover the program to be a danger, the e-taxpayer driven organizations vacillated. Contend that this disappointment was because of an assortment of Critical Failure Factors which ends with a straightforward manageability disappointment demonstrate. In outline, it is suggested that the e-government program neglected to be politically and institutionally supportable because of individuals, the board, social, and basic components (Helbig et al., 2009).

1.6 TRUST BASED SENTIMENT ANALYSIS (SA)

Sentiment analysis (SA) tries to comprehend a subject's frame of mind or emotional response toward a particular point (or brand). Notion examination utilizes specific devices, systems, and techniques to comprehend what individuals feel about an issue.

Opinion examination doesn't need to be confounded and specialized. It could be something as straightforward as inspiring an individual in a group to discover what is being said about your image via web-based networking media and recognize its value. There is no requirement of a major spending plan and speculation into confused programming (Zhang et al., 2014).

Figure 1.8 represents sentiment examination standards which can be applied to:

- Reviews and reactions of the product;
- Associations and Individuals;
- Training, themes, occasions, and issues.

On the web or internet-based life material including Facebook posts, web recordings, blog remarks, discussion comments, Tweets, recordings, designs, or pictures.

Instruments, strategies or techniques utilized in sentiment investigation shift broadly and may result in the integration of social media monitoring, biometric tools, computational linguistics, keyword/text processing tools, natural language processing (NLP) tools, or a basic appraisal by someone else (Zhang et al., 2014).

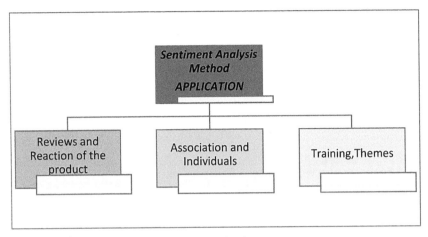

FIGURE 1.8 Sentiment analysis method application.

1.7   FORENSIC ANALYSIS IN WEB APPLICATIONS

These days, web applications are the prime target for security attackers. Utilizing explicit security systems can anticipate or identify a security assault on a web application, yet it is not possible to discover the criminal who has committed the security assault not being able to follow back an assault, encourage others to dispatch new assaults on a similar framework. Web application forensics intends to follow back and attribute a web application security assault to its originator. This may altogether diminish the security assaults focusing on a web application consistently, thus improving its security (Chen et al., 2013).

1.8   ONLINE TRUST EVALUATION

As presented in Figure 1.9 trusts is a critical foundation for access control in the field of online social organizations protection policies. In the present techniques, the subjectivity and individualization of the trust are over-looked and a fixed model is worked for every one of the clients. Truth be told, distinctive clients most likely take diverse trust highlighting into their contemplations when settling on trust choices. Moreover, in the present plans, just clients' static highlights are mapped into trust values, without the danger of security spillage. The protection spillage danger of the assessed client is evaluated through data stream forecasting. At that point, the User-Will and the protection spillage chance are altogether mapped into trust proof to be consolidated by an improved proof mix guideline of the proof hypothesis. At last, a few ordinary techniques and the proposed plan are executed to look at the execution on dataset opinions (Jiang et al., 2016).

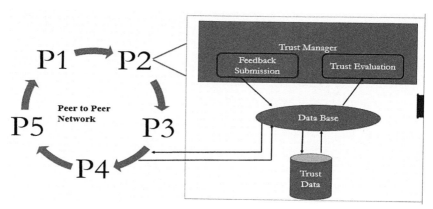

FIGURE 1.9   Online trusted system.

E-commerce sites have gained popularity. Many corporations are working together in terms of exploring the internet and capabilities. They are working on joint research for the same. Organizations like Amazon and booksellers have found ways to use the internet for their benefit (Sharma et al., 2012).

1.9   CONCLUSION

The statistics from various sources like worldometers, United Nations, IAMI, Facebook, and Cisco show that the country had 250 million active social media users in 2018. Figure 1.10 is the graphical representation is of the digital population in India.

By considering all the points and features mentioned above, it is concluded that the New Age Analytics technique has transformed the Internet and its services. It can also be said that today's techniques have grown the utility of the internet. The world of analytics and digital transformation appears limitless. As the internet is a huge source of data gathering, this gathered data is given some form by Analysis. Data

Analysis has become a popular term these days and has opened new career opportunities. Now it appears that this outburst of technology will just not stop here but it will reach more levels in different fields thus transforming the entire digital system.

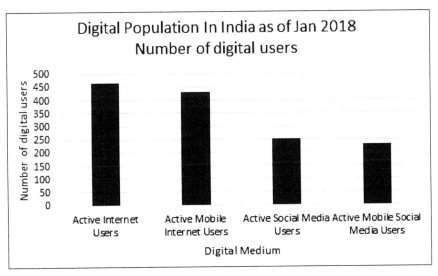

FIGURE 1.10   Illustration digital growth.

KEYWORDS

- **big data**
- **data analysis**
- **data mining**
- **digital marketing**
- **web analysis**

REFERENCES

Agrawal, R., Gupta, A., Prabhu, Y., & Varma, M., (2013). Multi-label learning with millions of labels: Recommending advertiser bid phrases for web pages. In: *Proceedings of the 22nd International Conference on World Wide Web* (pp. 13–24). ACM.

Bakos, Y., (2001). The emerging landscape for retail e-commerce. *Journal of Economic Perspectives, 15*(1), 69–80.

Banga, R., Bhardwaj, A., Peng, S. L., & Shrivastava, G., (2018). Authorship attribution for online social media. In: *Social Network Analytics for Contemporary Business Organizations* (pp. 141–165). IGI Global.

Bansal, H., Shrivastava, G., Nguyen, G. N., & Stanciu, L. M., (2018). *Social Network Analytics for Contemporary Business Organizations*. IGI Global.

Chen, Z., Han, F., Cao, J., Jiang, X., & Chen, S., (2013). Cloud computing-based forensic analysis for collaborative network security management system. *Tsinghua Science and Technology, 18*(1), 40–50.

Fan, W., & Gordon, M. D., (2014). The power of social media analytics. *Commun. Acm, 57*(6), 74–81.

Gupta, B. B., Sangaiah, A. K., Nedjah, N., Yamaguchi, S., Zhang, Z., & Sheng, M., (2018). Recent research in computational intelligence paradigms into security and privacy for online social networks (OSNs).

Helbing, N., Gil-García, J. R., & Ferro, E., (2009). Understanding the complexity of electronic government: Implications from the digital divide literature. *Government Information Quarterly, 26*(1), 89–97.

Hinchcliffe, D., (2018). *On Digital Strategy*. Retrieved from: dionhinchcliffe.com: https://dionhinchcliffe.com/category/digital-transformation/ (accessed on 16 February 2020).

Hinchcliffe, D., (2018). *What Enterprises will Focus on Digital Transformation in 2018?* Retrieved from: zdnet.com: https://www.zdnet.com/article/what-enterprises-will-focus-on-for-digital-transformation-in-2018/ (accessed on 16 February 2020).

Järvinen, J., & Karjaluoto, H., (2015). The use of web analytics for digital marketing performance measurement. *Industrial Marketing Management, 50*, 117–127.

Jiang, W., Wang, G., Bhuiyan, M. Z. A., & Wu, J., (2016). Understanding graph-based trust evaluation in online social networks: Methodologies and challenges. *ACM Computing Surveys (CSUR), 49*(1), 10.

Karim, M., & Rahman, R. M., (2013). Decision tree and naive Bayes algorithm for classification and generation of actionable knowledge for direct marketing. *Journal of Software Engineering and Applications, 6*(04), 196.

Konstas, I., Stathopoulos, V., & Jose, J. M., (2009). On social networks and collaborative recommendation. In: *Proceedings of the 32nd International ACM SIGIR Conference on Research and Development in Information Retrieval* (pp. 195–202). ACM.

Luo, X., Luo, H., & Chang, X., (2015). Online optimization of collaborative web service QoS prediction based on approximate dynamic programming. *International Journal of Distributed Sensor Networks, 11*(8), 452–492.

Persaud, A., & Azhar, I., (2012). Innovative mobile marketing via smartphones: Are consumers ready?. *Marketing Intelligence and Planning, 30*(4), 418–443.

Provost, F., & Fawcett, T., (2013). Data science and its relationship to big data and data-driven decision making. *Big Data, 1*(1), 51–59.

Robert, S. A. L., (2015). *Big Data Solution and the Internet of Things*. In: A. L. Robert.

Ruhi, U., (2014). Social media analytics as a business intelligence practice: Current landscape and future prospects. In: Umar, R., (ed.), *"Social Media Analytics as a Business Intelligence Practice: Current Landscape and Future Prospects," Journal of Internet Social Networking and Virtual Communities*.

Sharma, K., Shrivastava, G., & Singh, D., (2012). Risk impact of electronic-commerce mining: A technical overview. *Risk*, (29).

Shrivastava, G., Sharma, K., & Bawankan, A., (2012). A new framework semantic web technology-based e-learning. In: *2012 11ᵗʰ International Conference on Environment and Electrical Engineering* (pp. 1017–1021). IEEE.

Smith, K. T., (2011). Digital marketing strategies that Millennials find appealing, motivating, or just annoying. *Journal of Strategic Marketing*, *19*(6), 489–499.

Stackowiak, R., Licht, A., Mantha, V., & Nagode, L. (2015). *Big Data and the Internet of Things: enterprise information architecture for a new age*. Apress.

Truong, Y., & Simmons, G., (2010). Perceived intrusiveness in digital advertising: Strategic marketing implications. *Journal of Strategic Marketing*, *18*(3), 239–256.

Watson, C., McCarthy, J., & Rowley, J., (2013). Consumer attitudes towards mobile marketing in the smartphone era. *International Journal of Information Management*, *33*(5), 840–849.

Yadav, S., Ahmad, K., & Shekhar, J., (2011). Analysis of web mining applications and beneficial areas. *International Journal of IUM*, *12*(2).

Zhang, Y., Lai, G., Zhang, M., Zhang, Y., Liu, Y., & Ma, S., (2014). Explicit factor models for an explainable recommendation based on phrase-level sentiment analysis. In: *Proceedings of the 37ᵗʰ International ACM SIGIR Conference on Research and Development in Information Retrieval* (pp. 83–92). ACM.

this problem, the machines need to be trained to process and predict the future with the help of training and test data. To train the machine, various kinds of machine learning (ML) algorithms and tools are available under the category of classification, clustering, and prediction. It is essential to choose the significant algorithm and tool for the specific kind of data size, type, and its application domain. This chapter describes the various ML algorithms and tools in brief with the focus of its objective, working procedure, advantages, limitations, implementation tools, and the applicability of these algorithms in the real-time application domain.

2.1   INTRODUCTION

This section basically focuses on the various types of machine learning (ML) algorithms:

1.   **Supervised Machine Learning Algorithms:** In the supervised ML algorithm, the output variable (Y) is described using some functions on input variables (X).

$$Y = f(X) \tag{1}$$

Here, 'Y' is the test variable and 'X' is the training variable. Lots of algorithms are used which apply some functions on 'X' to build models to find 'Y.' The supervised ML algorithm is sub-divided into Classification for target label and Regression for quantitative prediction.

2.   **Un-Supervised Machine Learning Algorithms:** In this algorithm, the output variable (Y) is not available. Only input variables (X) are available without any specific class label.

$$Y = f(X) \tag{2}$$

Here, many algorithms are used which apply some functions on 'X' to make a group of labels based on some similarity to find 'Y.' The unsupervised ML is sub-divided into Clustering and Association.

3.   **Semi-Supervised Machine Learning Algorithms:** In semi-supervised ML algorithms, the input variables (X) are available with a mixture of labeled and un-labeled manner. The combination of supervised and unsupervised learning techniques is used to find 'Y.'

4. **Reinforcement Machine Learning Algorithms:** Based on reward/feedback the machines/software agents can take optimal decisions for a specific context to maximize the performance of a system.

This chapter mainly focuses on supervised and unsupervised ML algorithms and its relevant tools.

2.1.1 CLASSIFICATION

It is the process of finding the new data (test data) class label based on previously known (training data) class labels. The class labels are normally a category/qualitative variables like Student, Employee, Parents, etc.

Popular Classification Algorithms:

- Decision tree (DT);
- Naive Bayes (NB);
- K-nearest neighbors (KNN);
- Support vector machine (SVM).

2.1.2 REGRESSION

Like classification, regression is also a supervised learning algorithm. It finds the relationship between some independent (already known) variables with some dependent (unknown) variables. Rather than a categorical output value of classification, regression gives continuous quantitative output value like "price," "weight," "count" etc. based on already known input values. Some of the functions like quadratic, cubic, quadratic, power, logarithmic, and others are helpful to find the best approximation of regression value.

Popular Regression Algorithms:

- Linear regression;
- Logistic regression (LR).

2.1.3 CLUSTERING

In this technique, the un-labeled and scattered data points are grouped together based on some similarity. These groups are then labeled accordingly. The popular clustering algorithms are:

- K-means clustering; and
- Hierarchical algorithm.

The various supervised algorithms proposed in the literature are described below.

2.2   DECISION TREE (DT) ALGORITHM

2.2.1   OBJECTIVE

DT algorithm is mainly used to construct a training/classification/regression model in the form of a tree structure (root, branch, and leaf), which is based on (inferred from) previous data to classify/predict class or target variables of the future or new data with the help of decision rules or DTs.

2.2.2   WORKING PROCEDURE

The working procedure of the DT algorithm is described as follows.

The DTs are helpful for numerical and categorical data classification/prediction. Entropy and information gain (IG) based DTs are constructed using ID3 (Iterative Dichotomiser 3) algorithm and Gini Index based DTs are constructed using the CART (classification and regression trees) algorithm. It uses a greedy search methodology from the top-bottom without backtracking. In the complete DT, root node in each level is a starting point or the best splitting attribute in that position which helps to test on an attribute. The output of the test produces branches. Leaf nodes act as a final class label or target variable to classify/predict the new data. Classification rules are drawn from root to leaf node (Xia et al., 2018).

1. **Popular Attribute Selection Measures (Methods):** Finding the best attributes (attribute selection) to root node in each level from the given data set is a big challenging task in DT. The random selection gives bad results and low prediction accuracy.
   i.   Information gain → if attributes are categorical.
   ii. Gini index → if attributes are a continuous one.

2. **Information Gain (IG):** It helps to identify the feature of data set to be the best split of the tree at each level. IG is based on the concept of entropy.

3. **Entropy:** DT subsets may contain attributes with similar values (homogeneous). Entropy is used to measure/calculate the homogeneity/uncertainty/randomness of a random variable X. An entropy value ranges from 0–1, which is used to represent the entropy of information.

For a binary classification problem, only two classes are possible, that is: (1) positive (2) negative:

- The entropy will be 0, i.e., low, when all records are negative/positive.
- The entropy will be 1, i.e., high, when half of the records are negative/positive.
- If half of the records are positive (or) negative entropy will be 1, i.e., high.
- To calculate the IG, the entropy needs to be calculated.

$$Entropy(S) = \sum_{i=1}^{c} -p_i \log_2 p_i \qquad (3)$$

There are two steps for calculating IG:

i. Calculate the entropy of the target; and
ii. Calculate the entropy of every (each) attribute/branch.

$$IG = \text{Entropy of the target} - \text{Entropy of each attribute} \qquad (4)$$

IG value is sorted for each attribute and then the DT is constructed based on high IG value to low.

- If the IG value of the attribute is high, it will be placed as a root node.
- If branches (attribute) with entropy = 0, will be placed on the root node.
- If a branch (attribute) with entropy > 0, will be considered for further splitting at each level of root node.

4. **Gini Index:** The randomly selected incorrect element can be identified easily by using the Gini Index with lower value.

$$Gini = 1 - \sum_{i=1}^{c} (p_i)^2 \qquad (5)$$

5. **Over Fitting:** If DT construction gets deeper and deeper using training data, it will minimize the training set error, but maximizes the test set error. Also, a tree with many branches due to irregularities and outliers, data will lead to low prediction accuracy.

There are two approaches used to avoid overfitting:

i.   Pre Pruning: In this approach, DT branch expansion is stopped earlier before it gets a complex structure. In each level, if the root node's goodness value will be reached which is less than the threshold value, it is not recommended for the further splitting of the tree and to find the stopping point is a difficult task.

ii. Post Pruning: Using cross-validation data, the tree expansion of the root node in every level is done to verify whether it gives improvement or not. If it looks like an improvement one, the tree will be allowed to get expansion else stopped at that point and is labeled as a leaf node.

2.2.3 ADVANTAGES

The advantages of DT include the following:

- Easy to implement, explain, and understand;
- It can classify/predict categorical as well as numerical data;
- It takes less data preprocessing;
- Statistical test helps to validate the DT model;
- It can handle large data set very well;
- It resembles the human decision-making methodology;
- The complex tree structure can be easily understood by visualization.

2.2.4 LIMITATIONS

The limitations of DT include:

- Probability to occur overfitting in the DT is high.
- Basically, prediction accuracy is low compared to other ML algorithms.
- If class labels are huge, then the calculation may lead to complexity.
- The prediction result will not be good, if evaluation data and sampled training data are different.
- Outliers may produce sampling errors.
- Small change in the data set will lead a different tree structure and hence is not a stable one.
- Redrawing is needed for every addition of information to the data set.

2.2.5   IMPLEMENTATION TOOLS

The tools used for implementation of DT include:

- Weka;
- KNIME;
- Scikit-learn of python;
- R tool.

2.2.6   REAL-TIME APPLICATIONS

The algorithm can be applicable to the following identified real-time applications:

- Agriculture;
- Astronomy;
- Biomedical engineering;
- Control systems;
- Financial analysis;
- Text processing;
- Medicine.

2.3   SUPPORT VECTOR MACHINE (SVM)

2.3.1   OBJECTIVE

It is a type of supervised ML algorithm. It is a powerful tool to perform regression, classification, and outlier detection of data. The original inventor of the SVM is Vapnik & Chervonenkis. They proposed the SVM hyperplane with a linear classifier. Vapnik, Boser, & Guyon proposed SVM hyperplane with a non-linear classifier using the kernel concept. It can perform Support Vector Clustering for the unsupervised learning algo-rithm. The main objective is to find a hyperplane that divides the classes and the new data is classified based on the hyperplane.

2.3.2   WORKING PROCEDURE

The working procedure of SVM is described as follows.
    A model is built based on training data to classify the new unknown data. The hyperplane is drawn to separate data points based on classification.

Two thumb rules of drawing hyperplane are:

1. The hyperplane must completely separate the two classes in the best manner; and
2. The maximum-margin hyperplane should be chosen as the best separator.

The following two types of SVM classifiers are used:

1. Linear SVM classifier; and
2. Non-linear SVM classifier.

2.3.2.1   LINEAR SVM CLASSIFIER

Let us consider two-class training data that are plotted on the (x, y) graph or 2-dimensional space. Drawing a straight-line hyperplane separates each class data points. Now one class data points are on the left side of the hyperplane and others are on the right side. Also, draw a parallel straight line along the two sides of the hyperplane with maximum distance and touch the nearest data points of each class to the hyperplane (Pan et al., 2018). The touched data points of parallel lines are called support vectors. If support vectors are removed or moved somewhere, then the new hyperplane wants to find it out as per the procedure. Because, it helps to find the best division of hyperplane. The two parallel straight lines are called the maximum-margin hyperplane (maximum margin classifier). There may be many hyperplanes (1 to n dimensional) possible for segregating the data points, but the only maximum-margin hyperplane is chosen among them as a better division of two classes. It is called decision boundary. Linear SVM classifier comes under lower or 2-dimensional data space.

2.3.2.2   NON-LINEAR SVM CLASSIFIER

The higher dimensional training data point space is possible in the real world. For this kind of data, the straight-line hyperplane is not suitable. The best option is a nonlinear hyperplane with the kernel trick and is useful by drawing the maximum-margin hyperplane. The kernel trick may be the Polynomial kernel, Radial Basis function kernel, etc. Here, training data points are plotted on the higher dimensional space and kernel represents the distance between new data and support vectors. The different types of kernel used in the non-linear SVM classifier includes:

1. Polynomial (homogeneous) Kernel:

$$k(\vec{x}_i, \vec{x}_j) = (\vec{x}_i . \vec{x}_j)^d \tag{6}$$

2. Polynomial (non-homogeneous) Kernel:

$$K(x, y) = (x^T y + c)^d \tag{7}$$

3. Radial Basis Function Kernel:

$$K(x, x') = \exp(-\frac{\| x - x' \|^2}{2\sigma^2}) \tag{8}$$

4. Linear.
5. Sigmoid.

Outliers: In the real world, few data points of left side class in the hyperplane may be placed in the right side of the best hyperplane as well as few right-side class data points are placed in the left side of the hyperplane. These data points are called outliers and are omitted by SVM classifiers.

2.3.3 ADVANTAGES

The advantages of SVM include the following:

- This is suitable for high dimensional data spaces.
- It is effective, if the number of dimensions is greater than the number of samples.
- It supports nonlinear data classification using a kernel trick.
- It is a robust classifier for prediction problems.
- It is a memory-efficient classifier due to use of support vectors (subset of training points).

2.3.4 LIMITATIONS

The limitations of SVM include:

- The wrong kernel selection will increase the classification of error percentage.

- It is very slow in test phase even it has good generalization performance.
- It is highly complex from the algorithmic point of view.
- It requires huge memory space due to quadratic programming.
- It is not suitable for noise data sets and large data set due to its higher training time.
- It needs expensive five-fold cross validation (FCV) to calculate probability measures.

2.3.5   REAL-TIME APPLICATIONS

The algorithm may be best suited to the following real-time applications:

- Detection of face image;
- Categorizing the hypertext and text;
- Handwritten digit recognition;
- In the field of image classification;
- Bioinformatics field;
- Remote homology detection and the protein fold;
- Recognition of handwriting, environmental, and geosciences;
- Control of generalized prediction.

2.3.6   IMPLEMENTATION TOOLS

The popular tools used for implementation of SVM include:

- SVMlight with C;
- LibSVM with python, MATLAB or ruby;
- Weka;
- Scikit-learn;
- Weka.

2.4   K-NEAREST NEIGHBOR (KNN)

2.4.1   OBJECTIVE

KNN is a supervised classifier. It is a best choice for the classification kind of problems. To predict the target label of a new test data, KNN finds the distance of the nearest training data class labels with a new test data point in the presence of K value. Then counts the number of very closest data points using K value and concludes the new test data class label. To calculate the number of the nearest training data point's distance, KNN uses *K* variable value between 0 to 10 with the help of Euclidean distance (ED) function for continuous variables and Hamming distance function for categorical variables.

2.4.2 WORKING PROCEDURE

The working procedure of KNN algorithm is described as follows.

Let us consider the training data sample with *n* counts. Every data point x_i has associated class label c_i. Here, x denotes training data points and c denotes class labels. For the understanding purpose, the training data and its associated class is plotted in (x, y) graph. Also, the new test data point is placed in the same (x, y) graph to predict the class label. Then the distance between the test data point with all the training data points is calculated using any one of the distance functions as mentioned in the objective. The distance values are then arranged in descending order. Now using K variable value is used to count the number of training data points that are near to test data point. The class label of the maximum training data point within k value will be assigned to the class label of new test data (Park and Lee, 2018).

1. **Choosing the K Value:** The difficult part of the KNN algorithm is to choose the K value. The small K value influence to the noise in predicting the target class label and the biggest K value leads to overfitting probability. Also, the biggest K value increases the calculation time and reduces the execution speed. The formula $K = n^{(1/2)}$ is used to choose the K value. To optimize the test result, the cross-validation of data is performed on training data with different K values. An optimized value will be chosen based on the best accuracy for the test result.

2. **Condensed Nearest Neighbor:** It is the process of removing the unwanted data from the training data to increase the accuracy.

The steps for condensing the data include:

1. **Outliers:** Remove the abnormal distance data.

2. **Prototypes:** To find the non-outlier points, a minimum training set is used.

3. **Absorbed Points:** Used to identify non-outlier points correctly.

2.4.3 ADVANTAGES

The advantages of KNN include the following:

- It is strong enough if training data is large.
- It is simple and flexible with attributes and distance functions.
- It can support a multi-class data set.

2.4.4 LIMITATIONS

The limitations of KNN include:

- Finding suitable K value is a difficult task.
- It is difficult to choose the type of distance function and its implications for a specific data set.
- The computation cost is a little bit high due to find the distance between the test to all training data.
- This is a kind of lazy learner, it couldn't learn anything, only depends on K-nearest common class labels.
- Sometimes, change in K value will result in a change in the target class label.
- It requires large storage space.
- It needs large samples for high accuracy.

2.4.5 REAL-TIME APPLICATIONS

The algorithm is best suited to the following identified real-time applications:

- Text, handwriting mining;
- Agriculture;
- Finance;
- Medicine;
- Credit ratings;
- Image and video recognition.

2.4.6 IMPLEMENTATION TOOLS

The tools used for the implementation of KNN include:

- Weka;
- Scikit-learn of python;
- R tool.

2.5 NAIVE BAYES (NB) ALGORITHM

2.5.1 OBJECTIVE

The NB algorithm performs classification tasks in the field of ML. It can do classification very well on the data set even if it has huge records with multi-class and binary class classification problems. The main application of Naïve Bayes is text analysis and natural language processing (NLP).

2.5.2 WORKING PROCEDURE

The working procedure of the NB algorithm is described as follows.

Baye's theorem is required to understand (work with) the NB algorithm efficiently. Bayes theorem is used to combine the multiple classification algorithms to form NB classifier with a common principle (Wu et al., 2019).

1. **Bayes Theorem:** It works based on conditional probability. Conditional probability means, an event will occur with conditioned event already occurred. The formula to obtain conditional probability is:

$$P(A \mid B) = \frac{P(B \mid A)P(A)}{P(B)} \tag{9}$$

where:

P(A): Prior probability of an event A and A is not dependent on an event B in any way.
P(A|B): Conditional probability of an event A with conditioned on another event B. If an event A occurs, it should be dependent on event B which has already occurred.

P(B|A): Conditional probability of an event B with conditioned on the event A. If an event B occurs, it should be dependent on event A which has already occurred.

P(B): Prior probability of an event B. Here, B is not dependent on event A in any way.

2. **Naive Bayes Classifier:** This considers all the features (attributes) of the data set independently, which contribute to classify the new data even if the attributes have some dependency. It means, probability of one attribute should not impact the occurrence of probability of other attributes in the data set. Also, each attribute in the data set equally contributes to predict the new data class label. As per Bayes theorem, P(A|B) is called as posterior probability. In NB classifier, the posterior probability is calculated for all the attributes independently. Then the highest posterior probability attribute is taken as the most likely attribute and is called maximum A posteriori (MAP).

$$MAP(A) = \max(P(A|B)(10)$$

$$MAP(A) = \max(P(B|A)*P(A))/P(B) \tag{11}$$

Here, P(B) acts as an evidence probability with a constant value, which helps to normalize the result only. As P(B) is a constant, it can be omitted and it will not affect the MAP(A) value. So;

$$MAP(A) = \max(P(B|A)*P(A)) \tag{12}$$

Types of NB Algorithm

There are three Naïve Bayes algorithms.

- Gaussian Naive Bayes;
- Multinomial Naive Bayes; and
- Bernoulli Naive Bayes.
i. **Gaussian Naive Bayes:** If all the attribute values are continuous, the Gaussian NB classifier is useful. It performs normal distribution and calculates the mean and variance for all attribute values.
ii. **Multinomial Naive Bayes:** It is useful when attribute values are distributed in a multinomial form.
iii. **Bernoulli Naive Bayes:** This classifier is useful when attribute values are binary-valued.

2.5.3 ADVANTAGES

The advantages of the NB algorithm include the following:

- It performs quickly with high scalability and simple manner.
- It can be useful for continuous, binary, and multinomial distributed attribute values.
- It is a best choice for text classification.
- Understanding and model building becomes very simple for small and big data (BD) set.
- For irrelevant attributes, it is not sensitive.

2.5.4 LIMITATIONS

The limitations of the NB algorithm include:

- It cannot find the relationship among the attributes as it recognizes all the attributes are irrelevant.
- There is a possibility to occur "zero conditional probability problem," if the attribute class has zero frequency data items.
- The assumption of highly independence of attribute variables has not been possible in real life always.
- It is not suitable for regression problems.

2.5.5 REAL-TIME APPLICATIONS

The algorithm is best suited to the following identified real-time applications:

- Real-time prediction;
- Multiclass prediction;
- Text, emails, symbols, and name classification/spam filtering/sentiment analysis (SA);
- Recommendation system.

2.5.6 IMPLEMENTATION TOOLS

The tools used for the implementation of the Naïve Bayes algorithm include:

9. **Remove Collinearity:** To avoid overfitting due to highly correlated variables, perform pairwise correlation and remove which is most correlated.

10. **Gaussian Distributions:** To get a more reliable prediction, the Gaussian distribution of data is performed.

11. **Rescale Inputs:** The normalization or standardization operations for rescale of input variables are also used to improve the perdition reliability.

2.6.3 ADVANTAGES

The advantages of linear regression include the following:

- It shows the linear relationship between dependent and independent variables with optimal results.
- It is a simple model and is easy to understand.

2.6.4 LIMITATIONS

The limitations of Linear Regression include:

- It can predict only numeric output.
- It cannot explain properly what it has learned.
- It is not applicable for nonlinear data.
- It is very much sensitive with outliers.
- It requires that data must be independent.

2.6.5 REAL-TIME APPLICATIONS

The algorithm is best suited to the following identified real-time applications:

- Studying engine performance from test data in automobiles.
- OLS regression can be used in weather data analysis.
- Linear regression can be used in market research studies and customer survey results analysis.
- Linear regression is used in observational astronomy.
- Predictive analytic.
- Operational efficiency.

2.6.6 IMPLEMENTATION TOOLS

The tools used for implementation of Linear Regression include:

- R Language;
- Python;
- MATLAB.

2.7 LOGISTIC REGRESSION (LR)

2.7.1 OBJECTIVE

It is also a supervised learning (classification) algorithm which predicts future data based on previous data and is applicable to statistics and the ML fields mostly. It is based on categorical variables. It has two forms as (1) Binomial LR, and (2) multinomial logistic regression (MLR). Binomial LR works on binary form based on the likelihood of whether an event will occur or not basis. But Multinomial LR takes multiple values. Natural logarithm function based logistic function will help to decide whether the data points are fitted in the LR model or not. This function provides "S" shaped curve with limited 0 or 1 value. It is better than linear regression. It uses odds ratios to find the probability. It gives the nonlinear relationship between dependent (categorical) and independent (one or more) variables with the help of black box or softmax function. It is used in deep learning and neural network (NN) construction.

2.7.2 WORKING PROCEDURE

The working procedure of LR is described as follows.

It won't predict any numerical value rather it calculates the probability of finding input value belongs to which class. It follows binary classification, so it finds whether the given input belongs to that specific class is true or not (means a positive class or negative class) and the probability that the input belongs to that class. Odd ratio = (p/1-p). If an odd ratio's log value is positive, then the success probability will be greater than 50%.

The LR equation:

$$y = e^{\wedge}(b_0 + b_1 * x) / (1 + e^{\wedge}(b_0 + b_1 * x)) \tag{15}$$

where: y – predicted output; b0 – intercept term; b1 – coefficient of single input value (x).

It calculates logits score to predict the target class. Logits are nothing but the multiplication of activity score and weights of an event. Logits are given as input to softmax function to predict the target class probability. If the target class probability is high, then the Predicted target class is identified based on the highest target class probability (Dingeni et al., 2019).

1. **Softmax Function:**

$$\sigma(z)_j = \frac{e^{z_j}}{\sum_{k=1}^{k} e^{z_k}} \; for \; j = 1,...,k \tag{16}$$

This is used to calculate event probabilities. The output range varies from 0–1. The sum of the output will be equal to 1.

It is also named as a normalized exponential function. In this function, numerator calculates an e-power value of logits and the denominator calculates the sum of the e-power value of logits to find the probability. This function is used in NB Classifier, Multinomial Logistic Classifier, and Deep Learning.

2. **Multiplying Softmax Function:** Inputs (logits) with any value will give large value. So LR's predicted target class will be a high probability.

3. **Dividing the Softmax Function:** Inputs (logits) with any value will give small value. So LR's predicted target class will be less probability.

4. **Logistic Function (Sigmoid Function):** Helps to calculate population growth in the field of ecology.

It maps the real valued-number on its own "S-shaped" curve between the ranges 0 to 1. The formula for the logistic function is as given in Eqn. (17).

$$1/(1 + e^\wedge - value) \tag{17}$$

where, e: base on natural logarithms; value: numeric value to be transformed.

5. **Performance Calculation:** It measures the performance.

6. **Akaike Information Criteria (AIC):** Compares the similarity relationship between one data item to another and the minimum AIC value should be chosen to select the best-fit model.

7. **Null Deviance and Residual Deviance:** Null deviance predicts the response of the model for the best fit with nothing but with intercept. The null deviance with less value is the best.

    Residual deviance predicts the response of the model with the addition of independent variables. The residual deviance with less value is the best.

8. **Confusion Matrix:** The confusion matrix helps to plot the actual and predicted value of a model in the tabular form. It helps to measure the accuracy and overfitting of the model. The accuracy is calculated using the formula given in Eqn. (18).

$$Accuracy = \frac{True\,Positive + True\,Nagatives}{True\,Positive + True\,Negatives + False\,Positives + False\,Negatives} \quad (18)$$

The sensitivity and specificity help to draw the receiver operating characteristic (ROC) curve and the formula to compute these is given in Eqns. (7) and (8).

$$\left. \begin{aligned} True\,Negative\,Rate\,(\text{TNR}), specificity = \frac{A}{A+B} \\ False\,Positive\,Rate\,(\text{FPR}), 1 - specificity = \frac{B}{A+B} \end{aligned} \right\} sum\,to\,1 \quad (19)$$

$$\left. \begin{aligned} True\,Positive\,Rate\,(\text{TPR}), sesitivity = \frac{D}{C+D} \\ False\,Negative\,Rate\,(\text{FNR}) = \frac{C}{C+D} \end{aligned} \right\} sum\,to\,1 \quad (20)$$

9. **Receiver Operating Characteristic (ROC):** True positive and false positive rate is evaluated using the ROC. $P > 0.5$ is advisable for plotting the ROC for a more success rate. The area under the curve (AUC) is also called as an index of accuracy (A) helps to measure the performance perfection of the ROC. The high AUC value is considered as best. For the best ROC, True positive should be equal to 1 and false-positive should be equal to 0.

2.7.3   ADVANTAGES

The advantages of LR include the following:

•   It is robust;

- It does not have a linear relationship between the input variable and output variable;
- It works on nonlinear data;
- Good probability values are considered for measurements;
- The dependent variable and error terms are not distributed as a normal form;
- The dependent variable variances are different in each group.

2.7.4   LIMITATIONS

The limitations of LR include:

- It is difficult to identify the correct independent variables;
- It gives only binary output;
- The data points need to be independent;
- It gives over fitting values;
- It requires a large sample size to achieve stable results;
- It doesn't perform well, when feature space is too large.

2.7.5   REAL-TIME APPLICATIONS

The algorithm is best suited to the following identified real-time applications:

- Image segmentation and categorization;
- Geographic image processing;
- Handwriting recognition;
- Healthcare.

2.7.6   IMPLEMENTATION TOOLS

The tools used for the implementation of LR include:

- R language;
- Python;
- MATLAB.

In the following sections, the various unsupervised algorithms proposed in the literature are described.

2.8   K-MEANS CLUSTERING

2.8.1   OBJECTIVE

If the collected data of a particular concept are un-labeled, there should be a need to provide a label for that data. This labeling of data will help to understand the data very clearly and to make some useful decisions on that data. An unsupervised k-means clustering algorithm helps to do this. This algorithm groups the data based on similarities of the data feature and label it for each group. The number of groups (k value) needs to be decided initially, before labeling the group. To group the data set as per its similarities, this algorithm uses an ED measure as a key point.

2.8.2   WORKING PROCEDURE

The steps involved in K-means clustering algorithm is presented in this section.

1. **Assignment of Data Points to a Group:** This algorithm takes the collected data set as an input. Based on earlier mentioned k value, the initial centroid points (the center point of each group) are placed on random manner from the data set. Then each data point is taken and calculates the squared ED to each centroid point. Then the data point is assigned to a group which has less distance to the centroid point of that group. This process is continued until all the data points are assigned to a particular group (Alhawarat and Hegazi, 2018). The formula as given in equation (21) helps to do this task:

$$\arg\min_{c_i \in C} dist(c_i, x)^2 \qquad (21)$$

where: C = number of groups; c_i = centroid point of each group; x_i = data points of the data set; dist() = squared ED.

2. **An Update of the Centroid Point of a Group:** After assigning all the data points to a particular group, each group has a collection of data points. Then, each group data points mean distance is calculated to the centroid point and the initial centroid point position is now updated with a new one based on mean distance value. This process is iterated up to all the centroid points to get stable. This centroid point will help to assign a new data point to

an exact group which is nothing but clustering. The formula as given in Eqn. (22) helps to do this:

$$c_i = \frac{1}{|s_i|} \sum_{x_i \in s_i} x_i \tag{22}$$

where: c_i = centroid point of each group; x_i = data points of the data set.

2.8.3 ADVANTAGES

The advantages of K-mean clustering algorithm include the following:

- This is faster if data sets are large and k-value is small.
- The implementation of this algorithm is very simple.

2.8.4 LIMITATIONS

The limitations of K-mean clustering algorithm include:

- To identify the k-value is a risky task.
- It cannot perform well with varieties of data dimensions.
- It has problems to finalize the centroid point.

2.8.5 REAL-TIME APPLICATIONS

The algorithm is best suited to the following identified real-time applications:

- Academics;
- Search engine;
- Cancer identification;
- Wireless sensor network.

2.8.6 IMPLEMENTATION TOOLS

The tools used for implementation of K-means clustering include:

- Java;
- Apache spark;

- MATLAB;
- R.

2.9 HIERARCHICAL ALGORITHM

2.9.1 OBJECTIVE

The common clustering algorithm works on un-labeled data set. The hierarchical algorithm is also coming under the clustering-based algorithm. The un-labeled and non-grouped data set want to make some kinds of groups based on some similarities among them. The hierarchical algorithm works to achieve this objective. It can be implemented using either the Agglomerative or Divisive method (Liao and Liu, 2018).

2.9.2 WORKING PROCEDURE

The working procedure of a hierarchical algorithm using agglomerative and divisive method is given as follows:

1. **Agglomerative Method:** It follows the bottom-up approach. The scattered data set grouped together with some similarities to obtain the optimum number of groups. Some of its methods include:

   - Single linkage;
   - Complete linkage;
   - Average linkage;
   - Centroid distance;
   - Ward's method.

   Here, it takes:

   Time complexity = o(n3).
   Required memory = o(n2).

2. **Divisive Method:** The top-down approach is used in this method. The initial data set is considered as a single group. Then it is further subdivided into a number of groups with some similarities.

   A greedy algorithm is used for the above two methods to split/merge the data set. Also, ED is calculated in both the methods to merge/split the data set.

2.9.3   ADVANTAGES

The advantages of Hierarchical algorithm include the following:

- It is simple to implement;
- It is easy to understand.

2.9.4   LIMITATIONS

The limitations of Hierarchical algorithm include:

- It is slower even for less number of data.
- The reverse process is not possible.

2.9.5   REAL-TIME APPLICATIONS

The algorithm is best suited to the following identified real-time applications:

- Healthcare;
- Business.

2.9.6   IMPLEMENTATION TOOLS

The tools used for the implementation of Hierarchical algorithm include:

- SciPy;
- ALGLIB;
- MATLAB;
- Scikit-learn;
- R.
- Weka.

2.10   REAL-TIME APPLICATIONS: A CASE STUDY

A standard DT learning with the base learner is not an effective one while performing self-training on semi-supervised learning, because its probability estimation is not much reliable for prediction operation. But, the modified and improved techniques like the NB Tree, Laplace correction, a mixing of no-pruning, grafting, distance-based measure and Mahalanobis distance method for DT learner produces the best result of prediction with high confidence for both labeled and unlabeled data for self-training and semi-supervised learning (Tanha et al., 2017).

The KNN method along with random forest (RF) classifier gives the best text classification result of partially available information compared to other text classification methods. Also explicit semantic analysis (ESA) based method is not recommended for the text classification of the partial information document (Diallo et al., 2016).

The clients of the cloud storage outsourced their valuable data to different kinds of cloud storage providers for storage purpose with encrypted data by their own encryption key. This encryption method is based on the single key setting and works based on interactions between client and server frequently. The frequent client-server interaction on the multi-cloud environment leads to privacy leakage of their data. This privacy issue can be overcome by outsourced collaborative kNN protocol (OCkNN). The working procedure of OCkNN is based on clients' own key encryption, but frequent client-server interaction is not needed. By this, the client's valuable data are maintained as highly secured and with confidentiality (Rong et al., 2016).

Incorporating the approach of deep feature weighting in the NB for conditional probability estimates rather than the classification of NB with less independent feature weights will improve the NB classification accuracy compared to the standard NB approach and also it is not degrading the model quality much more. This concept is very much helpful in the text classification with the remarkable improvement of the huge amount of text data (Wang et al., 2016).

The new (or) future built-up expansion is simulated with densification and calibrated using MLR and validated using cellular automata (CA) of the Wallonia region (Belgium) of the development interval of (1990–2000) and the validation interval of (2000–2010) using unordered MLR and CA. This heterogeneous model considers neighboring land use cells and causative factors of the developmental stage. Also, simulated 2010 built up the pattern are evaluated with 2020 original built up a map using fuzzy set theory and genetic algorithm. This model provides overall good accuracy and is helpful for policymakers with uncertainties (Mustafa et al., 2018).

The large electronic medical data has lots of missing values, noise, and unbalanced classes which leads to a lack in the construction of predictive models. Also, common data mining methods give poor classification measures. But proposed multilevel weighted SVM (MLWSVM) performs the simultaneous classification of large data sets and iterated regression analysis. This method produces a more accurate classification of data with fast and improved computational time without loss of quality compared to standard available SVM of missing and noisy data collection (Razzaghi et al., 2016).

To find the zinc-binding sites in proteins the optimal model is constructed using a novel approach called Meta zinc Prediction based on sequence information and multiple linear regressions and the least square method were used for estimating the parameters. The AURPC predictor of the proposed approach achieves 90% on the Zhao data set. This result will increase 2–9% as compared to the other three predictors ZincExplorer, ZincFinder, and ZincPred. The accuracy and robustness of the proposed predictor are validated on non-redundant independent test data set. The AURPC value increases 2–8% as well as Precision, Specificity, and MCC values have increased to 5–8%, 2–8%, and 4–12%, respectively. The proposed predictor is better than the traditional one. Also, it is helpful for inferring protein function and treatment of some serious disease (Chen et al., 2017).

To improve the performance of radar detection, this paper proposes an intelligent constant false alarm rate detector based on the SVM in multiple kinds of background environments. To train the SVM, this approach uses the feature called variability index statistics. This detector correctly finds the threshold value of an appropriate detector for the current operational environment and classification results with intelligence. In a homogeneous environment, it produces low performance and it works well in nonhomogeneous environments with many targets and clutter edges (Hao et al., 2017).

In electronics instruments, for the degradation of internal circuits and in the fastest processes the White noise plays a dominant role. To overcome this problem, the Counter (a Frequency Counter) based on Linear Regression is used. Counter performs better than traditional and counters are with the smallest possible variance of 1.25dB. This approach is implemented based on SoC (Rubiola et al., 2016).

Developed a classification model to enhance the accuracy of the prediction of a diabetic patient using ML algorithms by preprocessing with the bootstrapping resample method. This experiment was conducted on the PIMA data set using the WEKA tool. In this work, five ML algorithms such as Naive Bayes, KNN (k = 1), KNN (k = 3), DT (J48), and DT (J48graft) were considered and analyzed the accuracy of each algorithm with and without preprocessing. The author of this paper concluded that as the results, all algorithms gave increased classification accuracy after performing pre-processing, but, DT outperforms with an increase in accuracy from 78.43% without preprocessing to 94.44% with proposed preprocessing technique (Khan and Ali Zia, 2017).

The authors performed a classification experiment for the prediction of onset diabetes using ML algorithms and their results were compared. The experiment was done on the PIMA data set with the WEKA 3.8 tool using NB, LR, multi-layer perceptron (MLP), SVM, IBK, Adaboostm1, bagging, OneR, DT (J48) and RF machine-learning algorithms. The author compared all algorithms based on accuracy, sensitivity, specificity, PPV, NPV, and AUC and concluded that LR performed best among all 10 classifiers with an accuracy of 78.01% (Aminul and Jahan, 2017).

The authors built a software widget to explain the predicted results of ML algorithms in an automatic manner. Normally ML models do not explain their prediction results. The demonstration is done on the electronic medical record data set from the Practice Fusion diabetes classification competition using the algorithms like combining/stacking eight boosted regression trees and four RFs using a generalized additive model with cubic splines for predicting type 2 diabetes diagnoses within the next year. The author concluded with a test result of an automatic explanation method and its prediction results were found to be 87.4% without degrading accuracy (Luo, 2016).

The author examined the ensemble classifiers like rough set balanced rule ensemble and fuzzy RF using Sant Joan de Reus University Hospital data set to verify the patient's diabetic retinopathy status. The fuzzy RFs produce 84% of accuracy with 80% of sensitivity and 85% of specificity as compared to other approaches (Saleh et al., 2018).

The authors developed an ensemble hybrid model to increase the performance and accuracy of the prediction of ML algorithms by merging the individual technique/method into one using base learner. In this article, four ML algorithms such as KNN, NB, RF, and J48 were taken and make an ensemble hybrid model. This model was applied on PIDD (Pima Indian diabetes data set) and the performance and accuracy of each algorithm was examined as well as an ensemble hybrid model was proposed and the performance was measured with the help of F-measure, Recall, and Precision. This experiment was done using WEKA 3.8.1, Java, and Netbeans

Alhawarat, M., & Hegazi, M., (2018). "Revisiting K-means and topic modeling, a comparison study to cluster Arabic documents." *IEEE Access, 6*, 42740–42749.

Ali, Z. U., & Khan, N., (2017). "Predicting diabetes in medical data sets using machine learning techniques." *International Journal of Scientific and Engineering Research, 8*(5), 1538–1551.

Aminul, M., & Jahan, N., (2017). "Prediction of onset diabetes using machine learning techniques." *International Journal of Computer Applications, 180*(5), 7–11.

Dingen, D., Van't Veer, M., Houthuizen, P., Mestrom, E. H., Korsten, E. H., Bouwman, A. R., & Van Wijk, J., (2019). "Regression explorer: Interactive exploration of logistic regression models with subgroup analysis." *IEEE Transactions on Visualization and Computer Graphics, 25*(1), 246–255.

Dramé, K., Mougin, F., & Diallo, G., (2016). "Large scale biomedical texts classification: A kNN and an ESA-based approaches." *Journal of Biomedical Semantics, 7*(1).

Fang, X., Xu, Y., Li, X., Lai, Z., Wong, W. K., & Fang, B., (2018). "Regularized label relaxation linear regression." *IEEE Transactions on Neural Networks and Learning Systems, 29*(4), 1006–1018.

Gnana, A., Leavline, E., & Baig, B., (2017). "Diabetes prediction using medical data." *Journal of Computational Intelligence in Bioinformatics, 10*, 1–8.

Jiang, L., Li, C., Wang, S., & Zhang, L., (2016). "Deep feature weighting for Naive Bayes and its application to text classification." *Engineering Applications of Artificial Intelligence, 52*, 26–39.

Jiang, L., Zhang, L., Li, C., & Wu, J., (2019). "A correlation-based feature weighting filter for Naive Bayes." *IEEE Transactions on Knowledge and Data Engineering, 31*(2), 201–213.

Li, H., Pi, D., Wu, Y., & Chen, C., (2017). "Integrative method based on linear regression for the prediction of zinc-binding sites in proteins." *IEEE Access, 5*, 14647–14657.

Liao, E., & Liu, C., (2018). "A hierarchical algorithm based on density peaks clustering and ant colony optimization for traveling salesman problem." *IEEE Access, 6*, 38921–38933.

López, B., Torrent-Fontbona, F., Viñas, R., & Fernández-Real, J. M., (2018). "Single nucleotide polymorphism relevance learning with random forests for type 2 diabetes risk prediction." *Artificial Intelligence in Medicine, 85*, 43–49.

Luo, G., (2016). "Automatically explaining machine learning prediction results: A demonstration on type 2 diabetes risk prediction." *Health Information Science and Systems, 4*(1).

Mustafa, A., Heppenstall, A., Omrani, H., Saadi, I., Cools, M., & Teller, J., (2018). "Modeling built-up expansion and densification with multinomial logistic regression, cellular automata, and genetic algorithm." *Computers, Environment and Urban Systems, 67*, 147–156.

Pan, X., Yang, Z., Xu, Y., & Wang, L., (2018). "Safe screening rules for accelerating twin support vector machine classification." *IEEE Transactions on Neural Networks and Learning Systems, 29*(5), 1876–1887.

Park, J., & Lee, D. H., (2018). "Privacy-preserving K-nearest neighbor for medical diagnosis in e-health cloud." *Journal of Healthcare Engineering*, 1–11.

Razzaghi, T., Roderick, O., Safro, I., & Marko, N., (2016). "Multilevel weighted support vector machine for classification on healthcare data with missing values." *PLoS One, 11*(5).

Rohul, J. M. A., (2017). "Analysis and prediction of diabetes diseases using machine learning algorithm: Ensemble approach." *International Research Journal of Engineering and Technology, 04*(10), 426–435.

Rong, H., Wang, H. M., Liu, J., & Xian, M., (2016). "Privacy-preserving K-nearest neighbor computation in multiple cloud environments." *IEEE Access*, *4*, pp. 9589–9603.

Rubiola, E., Lenczner, M., Bourgeois, P. Y., & Vernotte, F., (2016). "The Ω counter, a frequency counter based on the linear regression." *IEEE Transactions on Ultrasonics, Ferroelectrics, and Frequency Control*, *63*(7), 961–969.

Saleh, E., Laszczyński, J. B., Moreno, A., Valls, A., Romero-Aroca, P., De La Riva-Fernández, S., & Lowiński, R. S., (2018). "Learning ensemble classifiers for diabetic retinopathy assessment." *Artificial Intelligence in Medicine*, *85*, 50–63.

Tanha, J., Van Someren, M., & Afsarmanesh, H., (2017). "Semi-supervised self-training for decision tree classifiers." *International Journal of Machine Learning and Cybernetics*, *8*(1), 355–370.

Thiyagarajan, C., Anandha, K. K., & Bharathi, A., (2016). "A survey on diabetes mellitus prediction using machine learning techniques." *International Journal of Applied Engineering Research*, *11*(3), 1810–1814.

Verma, P., Kaur, I., & Kaur, J., (2016). "Review of diabetes detection by machine learning and data mining." *International Journal of Advance Research Ideas and Innovations in Technology*, *2*(3), 1–6.

Wang, L., Wang, D., & Hao, C., (2017). "Intelligent CFAR detector based on support vector machine." *IEEE Access*, *5*.

Xia, L., Huang, Q., & Wu, D., (2018). "Decision tree-based contextual location prediction from mobile device logs." *Mobile Information Systems*, 1–11.

CHAPTER 3

Machine Learning and Its Applicability in Networking

AMIT JAIN[1] and ANAND NAYYAR[2]

[1]*Associate Professor, University Institute of Computing, Chandigarh University, Chandigarh, India, E-mail: amit_jainci@yahoo.com*

[2]*Graduate School, Duy Tan University, Vietnam*

ABSTRACT

Machine learning (ML) plays an all-important role in applications that involve problem-solving and enabling automation in different areas. The main reason behind this turnaround is the voluminous amount of data availability, vast modifications in ML techniques along with development in computational strategies as a whole. Out of all these, ML is enormously having key importance in several common and complex issues coming up in operations and management in networking. As of date, several studies have been undertaken in relation to various upcoming fields of research in networking or for particular technologies related to it. In this chapter, quite a few applications of ML have been discussed in relation to important fields of networking in association with emerging technologies in it. The readers of this chapter will be benefited from a novel insight into the concepts of web network analysis by going through several principle and theories of ML and their application in networking. This includes traffic prediction, congestion control, routing, and classification, resource management, and fault management along with network security as major concepts of learning. Also, this study will focus on the challenges in research and provide insight into future aspects to look into rising prospects of applying ML in networking. Hence, this will act as an apt involvement towards the issues and challenges of ml in networking in relation to the analysis of web-based networking. This will enable carrying forward the study to new heights by studying the several aspects related to automated operations in networking and their related management.

3.1 INTRODUCTION

Machine learning (ML) aids in the scrutiny of the data and retrieve the knowledge. ML extends ahead of simple learning or mining knowledge, so as to utilize and improve knowledge with moment and experience. In nutshell, the objective is to recognize and view concealed patterns in "training" data. Discovered patterns are utilized for scrutinizing unidentified data, so as to group it collectively or map it to the identified groups. As a result, it initiates a change among the conventional programming model, where code is manuscript for automating the processes. ML builds the prototype that reveals the pattern in the data. Earlier in the past, the techniques in ML were firm and inept of abiding any distinction from the training data.

With advancements in ML, these techniques have become bendable and durable in their usage to several real-world situations, mostly from complex to commonplace fields. For example, ML plays a great role in the health care area like medical imaging (MRI, CT-scan, etc.) and computer-aided diagnosis. Most often, the tools which are laid upon the basic concepts of ML are used as their components. For instance, various search engines widely utilize ML for non-trivial jobs, like spell corrections, query suggestions, page ranking, and web indexing. As a result, the automation related to several facets of living, starting from home to autonomous vehicles, relates to the various techniques of ML. Thus, it has converged all together as a vital aspect in several systems which help in analysis, automation, and decision making.

Besides several advancements in ML techniques, there are several causes that play an important role in its restoration. The success of several methodologies in ML depends upon the voluminous data. Of course, there is a gigantic size of data in today's networks, that is expected to enhance multiple times with advancements in networks, like the internet of things (IoT) and its countless linked nodes (Gartner Inc., 2015). This promotes the implementation of ML which is not limited to the identification of hidden files and unexpected patterns, but is also applicable for learning and comprehending the procedures that create the datasets.

Till now, the advancements in computation provide storage along with its processing which acts as an essential means in training, validation, testing the ML prototypes for the massive volume of data. In general, Cloud Computing supports almost endless computation along with storage resources, whereas graphics processing units (GPUs) and tensor processing units (TPUs) (NVIDIA, 2017; Google Cloud, 2017) give faster training mechanism as well as association with a large amount of data. It is significant to specify here that a trained prototype of ML is utilized for implication on low-ended competent equipments, e.g., smartphones. In spite of such advances, still operations in network and their management are bulky, and several faults occur in network because of human intervention at large level (UCI, 2005). Faults in network, results in, increasing the monetary burden along with condemnation of network service providers. Thus, it relates to a colossal concern in forming autonomic flexible networks (Ayoubi et al., 2017).

As, there is earnest requirement for cognitive control in relation to operation and their management in the network, so it brings forward inimitable collection of confront for ML (Ayoubi et al., 2017). Also, every network is exclusive and further, there exist lack of adoption of principles to obtain equality among networks. For instance, the architecture of the network from an organization to organization is dissimilar to a large extent, so the patterns which are established to toil in a network might not be rational for a different network of the similar sort. Also, the network is persistently growing and the architecture restrains from application of a predetermined collection of patterns that assist in operations and management in the network. It is more or less not possible to manually remain confined to network administration, as it relates to constant escalation in the utilization of the data passing in the network along with the variety of equipments through the network.

3.2 ML IN NETWORKING

Earlier, Arthur Samuel (1959) introduced "ML," as *the field of study that gives computers the ability to learn without being explicitly programmed* (Puget, 2016). ML is widely used in four types of problems, namely, classification, regression, rule extraction, and clustering (Brownlee, 2013).

Techniques of ML are purposeful in varied problem areas. A strongly connected domain comprises of data analysis intended in support of hefty collection of data, called data mining (Alpaydın, 2014). Though, the techniques of ML are valuable within data mining, the idea is to decisively and thoroughly explore data—its characteristics, invariants, probability distributions, variables, temporal granularity, and their transformations. Nevertheless, ML extends ahead of data mining to forecast actions or series of actions. Usually, ML is best for deducing resolution to problems which has a huge representative dataset.

Further, as shown in Figure 3.1, techniques of ML are intended to recognize and utilize unknown sample of data:

i.   Relating the final analysis as collection of similar data;
ii.   Forecasting the conclusion of prospect actions;
iii. Analyzing the results of a series of data sets for extracting the rule.

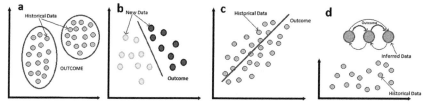

FIGURE 3.1   Beneficiaries of ML: a) clustering b) classification c) regression d) extracting rule. Reprinted with permission from Boutaba et al., 2018. (http://creativecommons.org/licenses/by/4.0/)

Figure 3.1 shows the statistics in a 2-D plane, for different datasets and their related outcomes. For example, in clustering, the result can be related to non-linear data representation using a hyperplane which distinguishes among the datasets. Whereas a problem of classification in networking is related to foresee the type of attack on the security as-denial-of-service (DoS), User-to-Root (U2R), Root-to-Local (R2L), in the available architecture of a network. In regression, predicament is evaluated to forecast in relation to emerging of collapse. Though there are diverse categories of problems that benefit from ML, as a common phenomenon is adapted to build the solutions.

Figure 3.2, shows the major components in constructing solution in networking using ML. Here, *Data* collection relates to assembling, creating, and explaining the relationship among the data and the associated collection of classes. *Feature Engineering* plays a vital role in reducing the data dimensionality and recognizes different characteristics which decrease computational burden, thereby increasing the accuracy of the outcome.

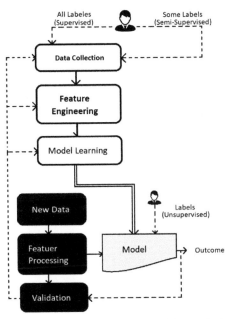

FIGURE 3.2 Components on ML model. Reprinted with permission from Boutaba et al., 2018. (http://creativecommons.org/licenses/by/4.0/)

3.2.1 PARADIGMS IN ML

In general, ML comprises of four key paradigms-*supervised learning, unsupervised learning, semi-supervised learning,* and *reinforcement learning* (RL). The categorical learning's persuade collection of data, feature engineering along with the procedure to set up ground truth. As the goal is to predict a result by utilizing the given dataset, so a prototype is designed which relates to the training dataset and associated labels of the available data. The final prediction is inferred as the recognition of belongingness to a collection of classes. In general, there exists dual thinking in relation to learning: creative and deductive (Pearl, 2014). The foundation of learning concepts is related to Bayes Theorem as stated in classification problem.

The general form of Bayes Theorem is (Boutaba et al., 2018):

$$P(A \mid B) = P(B \mid A) \times P(A)/P(B)$$

In Supervised learning, a prototype is build using training datasets. There exist several methodologies for labeling datasets. Supervised learning aids in identifying patterns among the datasets. In general, this technique is used in discrete-valued data (Classification) and continuous-valued data (Regression) problems. On the contrary, semi-supervised learning is useful in partial datasets. These are the one, which have partial labels available in relation to training data. In unsupervised learning, mixed data is used to identify the patterns among the datasets. It is widely applicable in clustering. For instance, issues related to density or outliers can be studied as a part of security attacks in networking as a similarity feature in ML.

RL refers to an agent-based calling procedure in relation to prototype building for making decisions. In general, in RL an agent establishes communication with the procedure of recognizing the patterns. At times, it is able to explore the knowledge, for which it is awarded or penalized accordingly. RL is widely applicable in the field of planning, scheduling, and decision making (Tesauro, 2007).

3.2.2   DATA COLLECTION

In ML, unbiased dataset is a prime requirement for developing an efficient prototype in similarity with networking issue. So it is a vital to collect the data invariably, as it differs from one scenario to another or from duration to duration. In general, the collection of data is performed in dual manner-either offline or online (Wang, 2018). In the case of offline, the data can be collected from pre-existing resources for utilization in training the model. On the contrary, online date relates to data collected from various surveys, feedbacks which can be utilized in training the prototype. There are certain other resources in relation to networking issues to be analyzed. Some theses resources include UCI knowledge discovery in databases (KDD) archive (UCI KDD, 2005), information marketplace for policy and analysis of cyber-risk & trust (IMPACT) Archive, etc. (IMPACT Cyber Trust, 2017; Sharma et al., 2018).

Another efficient methodology to gather data either offline or online is in utilization of tools related to measurement and monitoring of data. With the help of these tools several aspects related to networking like sampling rate, location, and duration of monitoring can be managed. In particular, various protocols meant for monitoring of network like simple network management protocol (SNMP), IP flow information export (IPFIX) can be utilized in monitoring (Harrington et al., 2002; Claise, 2008). Further, there exist active as well as passive monitoring (Fraleigh, 2001). In the case of active monitoring, probing is being performed in the network to collect required data from the network traffic. In contrary, the data is collected by learning from the behavior of actual data across the network. After collection, the data is splitted into various parts involving training, validation, and testing. On the basis of training dataset, the interconnectivity between the nodes is adjudged for designing a ML prototype. Further, the validation aids in selecting a framework from the available set of guidelines. In the last, testing of dataset is being performed to check the performance level of the designed prototype.

3.2.3 FEATURE ENGINEERING

At times, the dataset contains inseparable data which may be difficult to classify. So there is a requirement to refine the data, before going in for developing a prototype. This step elates to extracting the feature from the dataset, so as to differentiate among the class of interest. This has similarity in web-based concept in terms of level of granularity in networking.

Feature engineering is a vital step in ML as it involves the concept of selecting a feature and extracting it using a learning algorithm. It is widely utilized in dimensionality reduction, so as to identify diverse features to decrease the burden involved in computation along with increasing the efficiency of the prototype. In general, the concept of selecting the feature is related to pruning the irrelevant features from the prototype, whereas feature extraction relates to adding new features from the existing features. There are several techniques available like entropy, principal component analysis (PCA), etc. Some of the instances for selecting the feature involve traffic activity graphs (TAG) (Jin, 2010), backward greedy feature selection (BGFS), etc. (Erman et al., 2007).

3.2.4 ESTABLISHING GROUND TRUTH

It relates to assigning labels to the classes during prototype building. There exist several methodologies for giving names to datasets by utilizing the class features. For example, in traffic classification, pattern matching based on application signature is utilized to establish ground truth. In general, the application signatures are developed by making use of features like flow duration, average packet size, bytes per flow, etc. (Roughan, 2004a). These features act as a good differentiator (Roughan, 2004a). At times, the application signature can easily mine the signatures using unencrypted formats. Still, there is a need to keep these application signatures updated on regular basis (Haffner et al., 2005).

3.3   APPLICATIONS OF MACHINE LEARNING (ML) BASED TECHNIQUES

3.3.1   *TRAFFIC PREDICTION*

Forecasting the traffic in the network has a vital role to play in managing the operations in the network all together. It is basically related with the prediction of the network traffic and involves series of aspects as a part of time series forecasting (TSF). In TSF, the main goal is to develop a prototype using regression, for making a prediction in relation to existing network traffic as against the future network traffic.

3.3.1.1   *TIME-SERIES FORECASTING*

Traffic prediction is applied using multi-layer perceptron (MLP) in neural networks (NNs). The basic goal is related to improving the prediction in comparison to existing methodologies. In SNMP traffic, the collection of datasets is performed using separate ISP's along with resilient backpropagation in training the dataset. In it, the movement of packets is represented by the data across the network. Also, the movement of all the data packets is shown at the location of ISP. One of the mathematical techniques namely, linear interpolation aids in finding the unknown data in SNMP. Finally, NN ensemble is tested in different time series-dynamic forecasting, short-term forecasting, and long-term forecasting.

3.3.1.2   *NON-TIME SERIES FORECASTING (TSF)*

In comparison to the earlier forecasting technique, prediction can be made in relation to the network traffic by utilizing several methodologies and features. As mentioned, a recurrence based technique for the movement across the network, instead of focusing on amount of packets across the network (Li, 2016b). The main emphasis was on forecasting the movement of traffic packets inwards and outwards in the network. The prototype supported FNN which builds on backpropagation by making use of gradient descent & wavelet transform for capturing the recurrence and duration of packets across the network. The unknown data was filled using the computations made on the flow of packets and their data.

3.3.2 TRAFFIC CLASSIFICATION

It is a perfect for operating in the network, so as to be able to do varied functionalities along with their management. It consists of several aspects like performance monitoring, capacity planning, service differentiation, etc. At times, the network operator may look for providing priority to the movement of packets in relation to a particular task; thereby identifying anonymous data packets for variance recognition. At times, workload description is utilized in creating capable resource plan which influence wide-ranging applicability feat & resource requirements.

In Traffic, classification there is need to precisely correlate network traffic with existing classifications. These may be related to various protocols like FTP, HTTP, P2P, etc., or even online web links related to YouTube, Netflix, etc.

In general, the methodology is categorized as packet payload, host behavior and flow features (Bakhshi et al., 2016).

3.3.2.1 CLASSIFICATION BASED ON PAYLOAD

This basically looks for application signatures in the available payload and usually, has a steep costing in relation to storage and computation. Further, at times, problems are there in relation to keeping the record and adapting to the available signatures with volume of data across the network (Erman et al., 2007). Also, the privacy along with the security aspects ensures encryption of payload, so as to deny any unauthorized access. Thus, it becomes cumbersome to refer an application signature in relation to payload (Bernaille et al., 2007).

In a work, the researcher has decreased the computational cost involved by utilizing the small size of unidirectional data along with simple TCP flows in the form of feature vectors (Haffner et al., 2005). In relation to the protocols like SSH or HTTPS the encrypted data packets mine the characteristics from the simple procedure by negotiating the parameters of TCP connection.

In general, the method which depends upon controlling the movement of data packets has become unconventional for connection with more speed, whenever sampling is implemented (Ma, 2006). Further this; shortcoming is being put forward for refining signatures from any arbitrary location in a flow (Finamore et al., 2010). As the application involve streaming stress on investigating the payload for mining the signature instead of using UDP data packets, so the extraction of signature can also be accomplished by implementing Chi-square ($\chi2$) test for estimating the existence of random data in the initial memory size of every packet, which is further decomposed into the collection of sequential bits in buffer. Further, the value of randomness is estimated on the basis of difference in the estimated result and predicted result. Finally, the signatures are utilized for training the dataset in relation to the hyperplane or SVM classifier to separate the tuples into various classes. In spite of all this, still classification based on payload is utilized to classify the data with better accuracy score and thereby, it firms the ground truth.

3.3.2.2   CLASSIFICATION BASED ON HOST BEHAVIOR

In this method, the classification is influenced by the inbuilt behavioral manner of the host in the related network for forecasting the set of classes. It surmount the restrictions of unauthorized or misclassified port numbers and secure payload, by changing the inspection criteria to the end of the related network and thus, checks the movement of data packets between the hosts. As these hyperplane are based on the fact that varied applications can produce several patterns in communication, so the precision of this type of classification is purely based on the availability of the system used in monitoring, because the estimated pattern of communication is altered by the circulating asymmetries around the core of the network (Karagiannis et al., 2005).

3.3.2.3   CLASSIFICATION BASED ON FLOW-FEATURE

The flow-feature based hyperplane has different perception in comparison to classification using the concept of payload and host behavior. In flow-feature, the procedure is related to looking back and reviewing the communication held in different sessions, thereby emphasizing on the combination of the data packets which have already passed through. In general, the term complete flow refers to a single direction movement of data packets in a network from one port with a particular IP Address to another port with a separate IP Address by utilizing any protocol (Cisco Systems, 2017).

3.3.2.4 TRAFFIC CLASSIFICATION BASED ON NFV AND SDN

With the advancement in networking, certain novel terms have come into existence like virtualization and software-defined network. They assist in providing flexibility along with the certain efficient methodology for it. As it is widely known that the outcome of hyperplanes differs on the basis of flow-feature, so they comprehend particular aspects related to usability of the protocols and their applications in the network. Thus, discovering the suitable collection of characteristics is needed to obtain effectiveness in classification.

The goal is to choose the most efficient ML hyperplane along with the flow-features in cost-effective manner, by influencing a regulator along the virtual network. Thus, equilibrium is achieved among the precision and speed of classification thoroughly. Although, the selection of features has a vital role in the functionality of the network classification, still it is important to estimate the best hyperplane and control the characteristics of a feature or flow protocol. Still the expenses incurred in mining vary in characteristics from one another. This stands firm in the implementation of hyperplane. Thus, it is vital to:

i. Recognize the criteria in relation to selecting a feature on the data plane or other.
ii. Possess a central analysis of the resources allocated in the network, thereby enabling selection of hyperplane.

The controller is accountable for upholding the ML prototypes from offline preparation, and picking the best hyperplane for classification. Further, it takes care of the load in relation to estimating the resources in a network. The controller omits the control data packets as a part of the features of TCP traffic.

3.3.3   TRAFFIC ROUTING

As it plays an essential role in choosing a route for data broadcasting. In general, the criteria for selecting the different policies and goals for operation lead to lessen the expenses in movement of data packets. Traffic routing demands different capacities for prototypes in ML. It may relate to settling the estimates in relation to multifaceted and active network architecture, thereby enhancing the capacity for training among the chosen path and the capacity to foresee the results of routing decision. RL deploys agents to find the adjacent surroundings, without any control.

3.4   PERFORMANCE MANAGEMENT ISSUES

3.4.1   CONGESTION CONTROL

Congestion control is an elementary concept in operations & is liable to enable the movement of the data packets in the network. This encompasses guarantee, in terms of stability in the network, equal utilization of the resource & agreeable data packet loss ratio. In general, the varied network schemas implement their own collection of congestion control phenomenon. One of the best-used methodologies in congestion control is executed in transmission control protocol (Allman et al., 2009). TCP limits the rate of sending the packet across the network, when congestion occurs. Queue management acts as a popular method to operate in the intermediary devices on the network as an alternative to TCP (Braden et al., 1998). Over a period of time, there are numerous advancements in congestion control in relation to WWW and emerging architectures in networking, namely named data networking. Apart from these advancements, there exist several lacking in field like congestion, packet loss, queue management, etc.

3.4.2   RESOURCE MANAGEMENT

The concept of resource management involves controlling the important know-how's in the network. It includes various resources like memory, central processing unit, switches, routers, etc. In general, novel service providers adopt a pre-determined volume of resources which meets a required state of service. Moreover, it is important to forecast the requirement of resources in relation to computation and proper utilization in the network. Thus, an elementary issue in resource management is forecasting the requirement and making the provision for availability of resources. In spite of wide applicability of ML in load forecasting and resource management, several issues exist in varied networks, not limited to wireless networks, ad hoc networks, etc. (Prevost et al., 2011; Ojha et al., 2017). There exist several issues in managing the resource, out of which the two challenges are categorized as admission control & resource allocation (Bansal et al., 2018).

3.4.2.1 ADMISSION CONTROL

It is not a direct method to resource management which requires demand forecasting. The goal in admission control is related to having maximum utilization of resources in the network. For instance, a novel appeal for estimating the network resources is commenced for a Voice-over internet protocol. So, in such condition, admission control utters in relation to the novel inward demand to be settled or discarded based on existing network resources. Prominently, a new demand produces income to the service provider. Nevertheless, it might debase the QoS in obtainable services because of shortage in resources and will necessarily breach SLA, inviting punishment and thrashing in income. Consequently, there exist looming trade-off in accommodating new-fangled requirements and maintaining QoS. It deals with this confront to exploit as many as requests accepted and served to network exclusive of breaching SLA.

3.4.2.2 RESOURCE ALLOCATION

Resource allocation relates to a decision-making problem which dynamically takes care of resources to capitalize on a continuing intent, like income or resource consumption. The fundamental issue in allocation of resource is related to adapting resources for continuing payback in the face of randomness. In practice, the determined methodologies in relation to allocation of resource lacks in matching the frequency and amount of requests in the network. On the other hand, allocation of resource is paradigm for showing the reward of ML that discover and look after resource availability in several ways.

3.4.3 FAULT MANAGEMENT

It engrosses discovery, separation, and alteration of an anomalous state of a network. This calls for managing the operations & their look after to comprise a systematic acquaintance of the complete network, for its policies and all the executing tasks. Thus, fault management has become area of importance in modern networks. Novel fault management is immediate and is thought to have recurring procedure of uncovering, localization, and lessening of faults. Firstly, fault discovery mutually compares a variety of network indication to resolve whether individual or additional network malfunctions or faults have happened. For instance, faults can transpire owing to abridged switch capacity, augmented rate of generating packets for a particular application, inoperative switch, and immobilized links (Baras et al., 1997). Consequently, the next stride in fault management is isolation of the original cause of the fault(s) that needs concentrating on the physical locality of the damaged hardware or software, in addition to formatting the cause for the error. And finally, fault alleviation aspires to mend or approve the network performance. On the contrary, fault forecast is practical and aspires to avoid faults or malfunctions in the upcoming scenario, by forecasting them and commencing easing measures to lessen performance deprivation. The technique based on ML has been projected to cater to these issues and endorse cognitive fault recovery in the field of fault forecasting, discovery, isolation of origin, and alleviation of the faults.

3.5 FUTURE RESEARCH CHALLENGES AND OPPORTUNITIES

Provocations and complications of ML techniques had been discussed in the previous working and triumphs of ML. So, ML depends upon the accessibility of data which is combined with enhanced and supple ML algorithms to interpret composite tangle. Future networks are foreseen as a volatile increase in traffic magnitudes and strapped hardware will jeopardize outpouring statistics. In summation, these abilities will enhance with decrease in CAPEX, OPEX, or customer expectation. Also, effort is required to accomplish the cut-throat domain, by acquiring well planned and reasonable deployment.

Several emerging technologies like SDN or multi-tenancy should be adopted to decrease CAPEX, doubling utilization of resource and dispensing. Identically, autonomic architecture team up with SDN is considered to lesser OPEX. The aforesaid know-how's will permit potential networks for executing an ample selection of wide services along with available use cases. This continues to uphold other aspects like high speed broadband, lower latency & trustworthy services, end-to-end communication, tangible internet, business applicability, independent medium, dynamic supervision, and their management.

3.5.1 NETWORK PERSPECTIVE

3.5.1.1 EXPENDITURE OF INFERENCE

The incisiveness to capture data is due to rise in monitoring, in the network. This increases the demand of the network auditing schemes that are both systematic and meticulous. Predefined set of auditing probes in the hardware limits its pliability in monitoring.

3.5.1.2 COST OF DEBUGGING AND ANALYSIS REPORTS

Due to extensive FPR that makes operational settings, ML for foible discovery has acquired vast application in networks. But FPRs dissipate the time to explore the untrue alarm and decrease the certainty and belief in IDS. Secondly, missing of analyzed outline on deviations is also a big reason for reduced use of these techniques detected anomalies (Iyengar et al., 2015).

3.5.1.3 ENTANGLEMENT SITUATIONS

Traffic speculations, codification, vanquish, and obstruction on intervening junction is critical as they take less time and computing resources to circumvent degradation in network conductance. This unmentioned prerequisite exists, basically, in device-dependent networks, like WANETs and IoT. The acquisition estimates for ML assessment indicates that it is not easy to acquire the intricacy of ML-based concerns. Different from conventional algorithms, the algorithms in ML depends on dimensions and grade and acquisition targets of the datasets.

3.5.1.4   ML-A NEW PROFILE IN WEB

To refine the reliability and QoE for users, advanced procedure (e.g., HTTP/2, SPDY, and QUIC) has been approached that overpowers various consequences of HTTP/1.1 (Barman, 2004; Belshe, 2012; Iyengar et al., 2015). For example, HTTP/2 offers payload encryption, multiplexing, and concurrency, resource prioritization, and server push. Though, the WEB applications over HTTP/2 enjoy the benefits of these enhancements, it further complicates noise organizing by organizing volatility in the data used for ML.

3.5.1.5   ML FOR BRIGHT NETWORK STRATEGIES

The unmatched level and grade of unreliability in future networks will boost the complication of traffic engineering lessons, like congestion check, traffic forecast, codification, and routing. Moreover, ML-based results have done auspicious contribution to correct many traffic engineering aggravation, still the time intricacy is required to be calculated with the predicted dynamics, size of data, quantity of setups & inflexible applications requirements. To achieve this, affirmed policy-oriented traffic engineering perception can be obtained where operators can thoroughly and speedily adopt it. Policy-based classification of the traffic by utilizing SDN has determined proper results in the handling of QoS necessities depending on operator-engineered strategies (Ng, 2015). Strengthening ML to necessitate in developing adaptive procedures in strategic traffic engineering models still needs exploration. So, it is the most appropriate technique is RL applicability in relation to the future networks.

3.5.2   OUTLOOK OF A SYSTEM

3.5.2.1   REQUIREMENT OF ADJUSTABLE AND PROGRESSIVE LEARNING

As network pretends to be dynamic in applicability, so some of the features like amount of data packets; architecture and encryption techniques are still important, as they contribute widely in unmatched way. So, there is imme-diate requirement to develop a prototype based on ML for such innovative concepts. Out of all this, several prototypes related to ML are used and showcased in offline mode. With the rise in volume of available data for training purpose, there is a requirement to develop a technique which can enable to predict the outcome with accuracy. Of course, step-wise learning has other requirements to fulfill. For instance, if RL is used for routing in software-defined network, then several test cases are needed, so that the prototype can congregate to matching the results with related mapping of the guidelines. When novel data packets are sent in the network, then the SDN controller manages in relation to searching the best routing policy, in order to manage the same along with the numerous test cases, so as to keep track of the alterations in the connections. So, there is a requirement of a model which makes use of dataset and prototype in tandem, so as to generate minimum execution time for it.

3.5.2.2 REQUIREMENT OF SUPPORT FOR LEARNING

In general, ML is a solution for incompatible breaches, also termed as redundancy attacks which emphasize on fog learning (Barreno, 2006). As an example, at the time of implementing ML-based technique as attack on recognition of data packets, can result in contradictory hit which can deceit the prototype into classifying malfunction by destroying the training data. Therefore, it is essential to built robust prototypes which are competent to find the threats being posed in it. One of the most popular works to mention was being carried out by Cleverhans (Lu et al., 2009), where he created a collection of all such instances which provide a great deal into the adverse findings. There is study in relation to creating training data which is utilized to develop prototypes in ML, competent to separate genuine collection of data from the affected ones, specifically in the field of pattern recognition. So, some kind of use cases needs to be generated in relation to unseen conditions for preparing prototypes which can easily classify and predict the outcome. Further, this type of learning requires a model which defends the training dataset from any kind of violation or disruption, so as to implement stringent privacy and confidentiality in mutual exchange of datasets.

3.5.2.3 ARCHITECTURES FOR ML-DRIVEN NETWORKING

As of now, the emergence of newer networking paradigms involves huge amount of data, like several data tracks in relation to movement of data packets, performance issues, etc. Even in high-speed network, there can exist millions of records per day in relation to movement of data packets. So this, requirement is being managed with the aid of compatible collection of hardware and software for carrying out several tasks like storing, implementing, and analyzing. Also, the networks which have less resources, limited devices will reap rich benefits through ML system based on cloud environment. This will provide a huge storage and processing capabilities along with reliable network in relation to complex arrangement. Further, the procedure of collection of data and its analysis will be easily managed by this end-to-end setup. Certain highly mechanized ML prototypes, like Caffe2Go & Tensor Flow Lite, will automatically facilitate end-to-end nodes to get around the cloud-based environment and develop prototype indigenously.

3.5.3 *KNOWLEDGE PERSPECTIVE*

3.5.3.1 *MISSING REAL-WORLD DATA*

There is dire need to re-assess the work being carried out in several areas like resource management, fault management etc. In general, artificial data-sets have become naïve, thus, they do not pretend the intricacies involved in the real-world. Although, it is not easy to get the real-world data because of various security measures being adopted in the networks to avoid breach. Moreover, another requirement in relation to establishment of ground truth is difficulty faced, in case of large volume of data packets moving across the network, thus making it complex to check. Still, by introducing the faults in the network, a large amount of data can be generated. This doesn't aid to produce the training dataset available for usage by a ML prototype (Lu et al., 2009). So, there is a requirement to deal with such restrictions, so that ML models are utilized in more constructive manner.

3.5.3.2 *LACK OF EVALUATION MEASURE*

During the course of study, it became visible that comparison among every networking arena is next to possible, this may be due to several reasons like following under-rated performance measures, training datasets (Dainotti et al., 2012). So there is utmost need to follow standardized parameters for performance evaluation, as well as developing methodologies in relation to the analytical study for comparing and contrasting several network-related tasks. To achieve this objective, several standard organizations/setups like internet engineering task force (IETF), plays a vital role by encouraging standardization of assessment measures, performance measures, & data sets with the help of requests for comments (RFCs).

3.5.3.3   NEED OF THEORETICAL BASE

As the calculation and data storage blockades related to applicability of ML in networking is not an issue anymore, so question arises, *what exactly is stopping ML in networking?* The answer lies in missing theoretical base that is acting as one of the hindrance for ML. Although, an open-ended, research work is undergoing in this direction, to realize the need (Mestres et al., 2017). One major issue is the missing expertise, which is required for ML along with the scarcity in involvement of individuals in it. Thus, more cross-domain researches are required in ML along with networking.

3.6   CONCLUSION

ML's application in several fields of networking is widely known. This chapter aids in all-inclusive acquaintance on the areas and applicability of ML in relation to network operation & management, with emphasis on traffic engineering, performance optimization (Shrivastava, 2018). A detailed discussion in relation to the viability and realism of the projected ML mechanism in fulfilling the issues related to the self-controlled operation and management of prospective network architecture has been laid down.

Obviously, potential networks resolve to sustain a volatile expansion in traffic capacity and linked devices, to offer extraordinary capabilities for retrieving and distributing information. The unparalleled level and measure of ambiguity will augment the intricacy of traffic engineering responsibilities, involving congestion control, traffic prediction, classification, and routing, along with disclosure to faults and resource management. Though the ML-based mechanism contains capable outcome to tackle numerous traffic engineering issues, their extensibility requirements are required to be appraised with the predicted amount of data, quantity of equipments and applications. At times, accessible ML-based methodologies for fault and resource management spotlight typically on individual and single-layer networks. To expand the fault and resource management structure for potential networks, adopting ML methodologies should be absolute or re-formulated; so as to get into explanation of multi-tenancy in multi-layer networks.

KEYWORDS

- machine learning
- multi-layer networks
- network security
- traffic classification
- traffic prediction
- traffic routing

REFERENCES

Allman, M., Paxson, V., & Blanton, E., (2009). TCP congestion control. *RFC 5681, Internet Engineering Task Force.* https://tools.ietf.org/html/rfc5681 (accessed on 16 February 2020).

Alpaydin, E., (2014). *Introduction to Machine Learning* (3rd edn.). Cambridge: MIT Press.

Ayoubi, S., Limam, N., Salahuddin, M. A., Shahriar, N., Boutaba, R., Estrada-Solano, F., & Caicedo, O. M., (2018). Machine learning for cognitive network management. *IEEE Communication Magazine, 1*(1), 1.

Bakhshi, T., & Ghita, B., (2016). On internet traffic classification: A two-phased machine learning approach. *J. Comput. Netw. Commun.*

Bansal, H., Shrivastava, G., Nguyen, G. N., & Stanciu, L. M., (2018). *Social Network Analytics for Contemporary Business Organizations.* IGI Global.

Baras, J. S., Ball, M., Gupta, S., Viswanathan, P., & Shah, P., (1997). Automated network fault management. In: *MILCOM 97 Proceedings* (pp. 1244–1250). IEEE.

Barman, D., & Matta, I., (2004). Model-based loss inference by TCP over heterogeneous networks. In: *Proceedings of WiOpt 2004 Modeling and Optimization in Mobile* (pp. 364–373). Ad Hoc and Wireless Networks. Cambridge.

Barreno, M., Nelson, B., Sears, R., Joseph, A. D., & Tygar, J. D., (2006). Can machine learning be secure? In: *Proceedings of the 2006 ACM Symposium on Information, Computer and Communications Security* (pp. 16–25). ACM, ASIACCS '06. New York: ACM.

Belshe, M., & Peon, R., (2012). *SPDY Protocol*. Tech. Rep., Network Working Group. https://tools.ietf.org/pdf/draft-mbelshe-httpbis-spdy-00.pdf (accessed on 16 February 2020).

Bernaille, L., & Teixeira, R., (2007). Implementation issues of early application identification. *Lect. Notes Compu. Sci., 4866*, 156.

Boutaba, R., Salahuddin, M.A., Limam, N. et al. A comprehensive survey on machine learning for networking: evolution, applications and research opportunities. J Internet Serv Appl 9, 16 (2018). https://doi.org/10.1186/s13174-018-0087-2. (accessed on 16 February 2020).

Braden, B., Clark, D., Crowcroft, J., Davie, B., Deering, S., Estrin, D., et al., (1998). Recommendations on queue management and congestion avoidance in the internet. *RFC 2309, Internet Engineering Task Force*. https://tools.ietf.org/html/rfc2309 (accessed on 16 February 2020).

Brownlee, J., (2013). *Practical Machine Learning Problems*. https://machinelearningmastery.com/practical-machine-learning-problems/ (accessed on 16 February 2020).

Cisco Systems, (2012). *Cisco IOS Netflow*. http://www.cisco.com/go/netflow (accessed on 16 February 2020).

Dainotti, A., Pescape, A., & Claffy, K. C., (2012). Issues and future directions in traffic classification. *IEEE Netw., 26*(1), 35–40.

Erman, J., Mahanti, A., Arlitt, M., & Williamson, C., (2007b). Identifying and discriminating between web and peer-to-peer traffic in the network core. In: *Proceedings of the 16th International Conference on World Wide Web* (pp. 883–892). ACM.

Erman, J., Mahanti, A., Arlitt, M., Cohen, I., & Williamson, C., (2007a). Offline/real time traffic classification using semi-supervised learning. *Perform Eval., 64*(9), 1194–213.

Finamore, A., Mellia, M., Meo, M., & Rossi, D., (2010). Kiss: Stochastic packet inspection classifier for UDP traffic. *IEEE/ACM Trans Netw., 18*(5), 1505–1515.

Fraleigh, W. C., Diot, C., Lyles, B., Moon, S., Owezarski, P., Papagiannaki, D., & Tobagi, F., (2001). Design and deployment of a passive monitoring infrastructure. In: *Tyrrhenian International Workshop on Digital Communications* (pp. 556–575). Springer.

Gartner Inc., (2017). *Gartner Says 8.4 Billion Connected "Things" will be in Use in 2017*. Up 31 Percent From 2016. https://www.gartner.com/newsroom/id/3598917 (accessed on 16 February 2020).

Google, (2017). Cloud TPUs - ML accelerators for tensor flow. *Google Cloud Platform*. https://cloud.google.com/tpu/ (accessed on 16 February 2020).

Haffner, T. P., Sen, S., Spatscheck, O., & Wang, D., (2005). ACAS: Automated construction of application signatures. In: *Proceedings of the 2005 ACM SIGCOMM Workshop on Mining Network Data* (pp. 197–202). New York: ACM.

IMPACT Cyber Trust, (2017). *Information Marketplace for Policy and Analysis of Cyber-Risk and Trust*. https://www.impactcybertrust.org (accessed on 16 February 2020).

Internet Engineering Task Force, (2002). *Architecture for Describing Simple Network Management Protocol (SNMP) Management Frameworks*. https://tools.ietf.org/html/rfc3411 (accessed on 16 February 2020).

Internet Engineering Task Force, (2008). *Specification of the IP Flow Information Export (IPFIX) Protocol for the Exchange of IP Traffic Flow Information*. https://tools.ietf.org/html/rfc5101 (accessed on 16 February 2020).

Iyengar, J., & Swett, I., (2015). *QUIC: A UDP-Based Secure and Reliable Transport for HTTP/2*. Tech. Rep. network working group. https://tools.ietf.org/pdf/draft-tsvwg-quic-protocol-00.pdf (accessed on 16 February 2020).

Jin, Y., Duffield, N., Haffner, P., Sen, S., & Zhang, Z. L., (2010). Inferring applications at the network layer using collective traffic statistics. In: *International Teletraffic Congress (ITC) IEEE* (22nd edn., pp. 1–8).

Karagiannis, T., Papagiannaki, K., & Faloutsos, M., (2005). BLINC: Multilevel traffic classification in the dark. *ACM SIGCOMM Comput. Commun. Rev., 35*(4), 229–240.

Khari, M., & Shrivastava, G., (2018). Recent research in network security analytics. *International Journal of Sensors Wireless Communications and Control, 8*(1), 2–4.

Li, Y., Liu, H., Yang, W., Hu, D., & Xu, W., (2016b). *Inter-Data-Center Network Traffic Prediction with Elephant Flows: IEEE*, pp. 206–213.

Lu, X., Wang, H., Zhou, R., & Ge, B., (2009). Using hessian locally linear embedding for autonomic failure prediction. In: *Nature and Biologically Inspired, Computing* (pp. 772–776). NaBIC, World Congress on. IEEE.

Ma, J., Levchenko, K., Kreibich, C., Savage, S., & Voelker, G. M., (2006). Unexpected means of protocol inference. In: *Proceedings of the 6th ACM SIGCOMM Conference on Internet Measurement* (pp. 313–326).

Mahmoud, Q., (2007). *Cognitive Networks: Towards Self-Aware Networks*. Wiley-Interscience.

Mestres, A., Rodriguez-Natal, A., Carner, J., Barlet-Ros, P., Alarcón, E., Solé, M., Muntés-Mulero, V., Meyer, D., Barkai, S., Hibbett, M. J., et al., (2017). Knowledge-defined networking. *ACM SIGCOMM Comput. Commun. Rev., 47*(3), 2–10.

Ng, B., Hayes, M., & Seah, W. K. G., (2015). Developing a traffic classification platform for enterprise networks with SDN: Experiences amp, lessons learned. In: *IFIP Networking Conference* (pp. 1–9).

NVIDIA, (2017). *Graphics Processing Unit (GPU)*. http://www.nvidia.com/object/gpu.html (accessed on 16 February 2020).

Ojha, R. P., Sanyal, G., Srivastava, P. K., & Sharma, K., (2017). Design and analysis of modified SIQRS model for performance study of wireless sensor network. *Scalable Computing: Practice and Experience, 18*(3), 229–242.

Papernot, N., Goodfellow, I., Sheatsley, R., Feinman, R., & McDaniel, P. C., (2016). *v1. 0.0: An Adversarial Machine Learning Library*. arXiv preprint arXiv:161000768.

Pearl, J., (2014). *Probabilistic Reasoning in Intelligent Systems: Networks of Plausible Inference*. Morgan Kaufmann.

Prevost, J. J., Nagothu, K., Kelley, B., & Jamshidi, M., (2011). Prediction of cloud data center networks loads using stochastic and neural models. *IEEE*, pp. 276–281.

Puget, J. F., (2016). *What is Machine Learning?* IT best kept secret is optimization. https://www.ibm.com/developerworks/community/blogs/jfp/entry/What_Is_Machine_Learning (accessed on 16 February 2020).

Roughan, Y. M., Sen, S., Spatscheck, O., & Duffield, N., (2004a). Class-of-service mapping for QoS: a statistical signature-based approach to IP traffic classification. In: *Proceedings of the 4th ACM SIGCOMM Conference on Internet Measurement* (pp. 135–148). ACM.

Sansanwal, K., Shrivastava, G., Anand, R., & Sharma, K., (2019). Big data analysis and compression for indoor air quality. *Handbook of IoT and Big Data, 1*.

Sharma, K., & Gupta, B. B., (2018). Mitigation and risk factor analysis of android applications. *Computers and Electrical Engineering, 71*, 416–430.

Shrivastava, G., & Bhatnagar, V., (2011). Analysis of algorithms and complexity for secure association rule mining of distributed level hierarchy in web. *International Journal of Advanced Research in Computer Science, 2*(4).

Shrivastava, G., (2018). Investigating new evolutions and research in digital forensic and optimization. *Recent Patents on Engineering, 12*(1), 3, 4.

Sommer, R., & Paxson, V., (2010). Outside the closed world: On using machine learning for network intrusion detection. In: *Security and Privacy (SP), IEEE Symposium on, IEEE* (pp. 305–316).

Tesauro G., (2007). Reinforcement learning in autonomic computing: A manifesto and case studies. *IEEE Internet Comput., 11*(1), 22–30.

UCI KDD Archive, (2005). https://kdd.ics.uci.edu/ (accessed on 16 February 2020).

Wang, M., Cui, Y., Wang, X., Xiao, S., & Jiang, J., (2018). Machine learning for networking: Workflow, advances and opportunities. *IEEE Networks, 32*(2), 92–99.

CHAPTER 4

Machine Learning to Gain Novel Insight on Web Network Analysis

MEENAKSHI SHARMA,[1] ANSHUL GARG,[2] and NITIN JAIN[3]

[1]Galgotias University, India

*[1]School of Computer Science & Engineering, Galgotias University, India,
E-mail: minnyk@gmail.com*

*[2]Department of Computer Applications, Chandigarh Group of Colleges,
Landran, Mohali, Punjab, India*

*[3]School of Computer Science and Engineering, Galgotias University,
Greater Noida, UP, India*

ABSTRACT

The web is one of the most predominant sources of information and knowledge mining as it contains a large amount of data and the data is also increasing day by day. The global use of Web leads to unprecedented development in many fields of real-life like education, social, commercial, etc. (DINUCĂ, 2011). For these purposes, visitors on websites use some methods to analyze the data on the web. To analysis, the Web, not only the content but the hyperlinks structure of the Web and its diversity also plays an important role (DINUCĂ, 2011). Such analysis can improve the user's efficiency and effectiveness in searching for the required information. Web network analysis is used to identify the inter-relationship and intra-relationship between actors. In today's scenario, Machine learning (ML) is constructive in retrieving the most related information required by the user. It finds the hidden basic structures and the hyperlinks between various nodes in the web network, which is very helpful in rapid and effective web searching.

4.1 INTRODUCTION

The web is a miniature of the world network. Data on the web is much cheap and fast. In today's world, business organizations are looking at the Web as a big market. They are using the web for enhancement of their business by carefully examining the behavior of the user on the Web. Web applications use ML techniques to understand user behavior and to boost the user engagements. Machine learning (ML) uses various algorithms to know the features and functionality expected by users. ML models can detect the patterns from the collected data and can decide the actions that are purely based on the detected patterns (Chen, 2011). For better understanding the user behavior not only the analysis of content and usage on Web is mandatory but also how data on the Web are linked with each other this understanding is also a must. Networking is the connectivity relationship between the nodes. To analysis the web network, we have first to understand the relationships and connections between the different actors, which lead to a better understanding of the evolution and configuration of this virtual world. This will give reliable and related information, which can be helpful in better decision-making. ML can make the system able to scrutinize the data. The primary goal is to explore and utilize the hidden patterns in the training data, which can be used to scrutinize the anonymous data. ML techniques are becoming an essential facet in numerous systems that are used in the analysis, decision-making, and automation.

4.2 WEB NETWORK ANALYSIS

Web network analysis is generally used to enhance the performance of the web site. Web network highlights on the linking patterns of web objects like a website, a web blog, a web service, a social site like Facebook, Twitter, YouTube, etc. or anything on the web which can act as a node. In the web network, the in-link term can be used to show the hyperlinks pointing to the page and out-link to show the no of links present on the page. In web network analysis, the recognition of most prominent, strong, and central actors, identification of hub and authorities and discovery of various communities, etc. are significant as they are beneficial in knowledge mining from networks and helpful in problem-solving. Theoretically, a web network can be represented as graph G =

{V, E} Where V is the set of nodes and E is the set of edges. Web network can be represented as a directed, weighted, and undirected graph. A web network contains only one type of nodes then it is called homogeneous otherwise heterogeneous. Basic terminologies used in the web network can be explained in Figure 4.1.

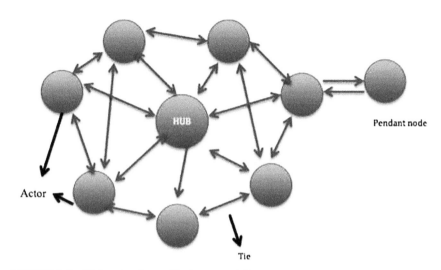

FIGURE 4.1 Web network graphical representation.

> The node also called actor or point can be anything like a social site, a web document, etc. In a network, when two nodes are directly connected with the edge, they are called neighbors. Two neighbors are called strongly connected if there is a directed path from each node to each other. Two neighbors are called weakly connected if there is a path from one node to another, but the orientation of an edge is not in consideration. A node is called pendant if it is tied only in one group with one connection. If only one actor gets attention for analysis, then it is called ego and group of actors those are ties with this ego called alters. A group of actors in which each actor chooses each other is called a reciprocated group. The degreedeg(x) of a node x in the directed graph is the sum of in-links and out-links of that node. The average degree can be calculated by taking the average of all the degrees of all actors present in that network.
> Tie also called link, edge, connection, or the relation between the two actors. It can be:

1. **Directed Tie:** Showing the flow of direction.
2. **Bounded Tie:** Simple and Undirected.
3. **Dichotomous Tie:** Present or absent between two actors.
4. **Weighted Tie:** Measured on a scale.

The path between two actors X and Y is the sequence of actors coming in between and the length of path d (X, Y) is calculated as the total minimum number of edges that are required to reach from actor X to actor Y also called the shortest path between two actors. The average path length in the web network is the arithmetic mean of the path between all the actors of the network. The diameter in the web network is the longest possible path in the network. Networks in which actors are connected are called valid networks.

To analyze the web network, there are some measures with are very helpful. These are discussed in subsections.

4.2.1 DENSITY OF NETWORK

Density is the cohesion in the network (Sagar et al., 2012). The density of the network is the present ratio edges in the graph to the maximum possible edges in the graph if the graph is complete. It can be calculated as:

$$\Delta = 2 \times n / g\,(g - 1)$$

Where 'n' is the no of edges present in the graph and 'g' is the no of the maximum number of possible edges is the graph is a complete graph (Plazzl et al., 2012; Sagar et al., 2012) (Figure 4.2).

4.2.2 CLUSTERING COEFFICIENT (CC)

It is the probability that two neighbor actors are connected with an edge.

$$cc_x = 2e_x / k_x(k_x - 1) \text{ (Plazzl et al., 2012; Sagar et al., 2012)}$$

where: cc_x = Clustering coefficient of an actor 'X'; e_x = Sum of all edges between the neighbors of actor 'X'; k_x = Number of all neighbors of actor 'X.'

The clustering coefficient (CC) should be $0 \leq cc_x = \leq 1$. If the CC is 0 it means the node is acting like star means there is no connection between the neighbors of node x but if the CC is 1, it means it is acting like a clique and all its neighbors are also connected with each other.

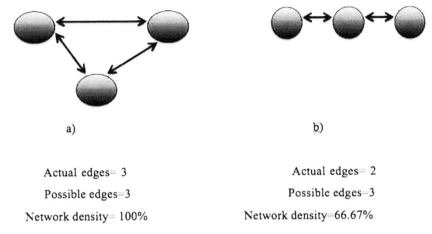

a)   b)

Actual edges= 3   Actual edges= 2
Possible edges=3   Possible edges=3
Network density= 100%   Network density=66.67%

FIGURE 4.2   Density of web network.

In Figure 4.3, the number of neighbors of node x (k_x) = 3(shown by a green color circle).

The edge between the neighbors = 1(shown by a red color edge)

So, the clustering coefficient (CC) = $2*(1)/(3*(3-1) = 1/6$

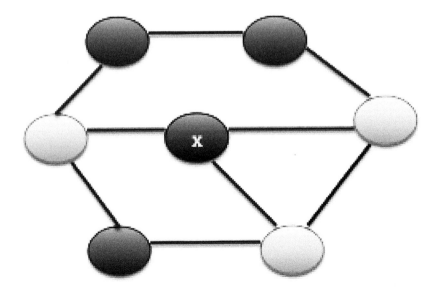

FIGURE 4.3   Clustering coefficient.

In this example 1/6 is the fraction of possible interconnections between the node x. The maximum no of connection node x can have are 6.

The CC of the whole graph G is the arithmetic mean of the CC of all the nodes in the network.

4.2.3   CENTRALITY MEASURE

This measure is used to identify the most critical node in the network. If a node plays an essential role as compared to other nodes in the network, it will act as a central node (Sagar et al., 2012). To measure the centrality, there are some criteria like shortest path, Random Process, Feedback, Current flow, Rehabilitee, etc. (Das et al., 2018). The centrality of a node can be measured in many terms.

4.2.3.1   DEGREE CENTRALITY

It is the number of edges that are connected to the central node (Sagar et al., 2012). More the degree centrality more critical the node is. The normalized degree centrality of a node can be calculated as:

$$C_{deg}(x) = \frac{deg_x}{N-1} \text{ (Das et al., 2018; Freeman, 1978)}$$

where deg_x is the degree of node x and N is the number of nodes in the network. E.g., If we calculate the degree centrality of node x in Figure 4.3 then it must be:

$$C_{deg}(x) = \frac{3}{7-1} = 3/6 = 0.5$$

The degree of node x is three and the number of nodes present in the network is seven. So the degree centrality is 0.5.

If the graph is directed than the in-degree centrality and out-degree centrality will be calculated differently with the formulas:

$$C_{indeg}(x) = \frac{deg_x^{in}}{N-1} \text{ And } C_{outdeg}(x) = \frac{deg_x^{out}}{N-1}$$

The degree centrality is generally used in the local structure as it can provide the information regarding how much edges are immediately related to the node.

4.2.3.2 CLOSENESS CENTRALITY

It is the reciprocal of the sum of all geodesic distances to every other node from each node (Plazzl, 2012; Sagar et al., 2012; Sabidussi, 1966).
It can be calculated as:

$$C_c(X) = \frac{1}{\sum_{X,Y \in V} d_{(X,Y)}}$$

where: $d_{x,y}$ is the distance between two nodes x and y.
The normalized closeness centrality can be calculated as:

$$C_c'(X) = \frac{N-1}{\sum_{X,Y \in V} d_{(X,Y)}}$$

where N is the no of nodes in the network.

4.2.3.3 BETWEENNESS CENTRALITY

Betweenness centrality can be calculated as:

$$C_B(x) = \sum_{y \neq x \in V} \sum_{z \neq x \in V} \sigma_{yz}(x) / \sigma_{yz}$$

where σ_{yz} represents total no of shortest paths between the node y and z and $\sigma_{yz}(x)$ represents the no of shortest paths that pass through the node x. The network having high betweenness can play a role of the bridge the a between the two network areas (Freeman, 1978; Plazzl et al., 2012; Sagar et al., 2012).
We can either include or exclude the node x while calculating the between centrality.
When X is excluded then:

$$C_B(X) = \frac{\sigma_{A,Y}(X)}{\sigma_{A,Y}} + \frac{\sigma_{A,Z}(X)}{\sigma_{A,Z}} + \frac{\sigma_{Z,Y}(X)}{\sigma_{Z,Y}}$$

$$C_B(X) = \frac{1}{1} + \frac{1}{1} + \frac{0}{1} = 2$$

When X is included then:

$$C_B(X) = \frac{\sigma_{A,X}(X)}{\sigma_{A,X}} + \frac{\sigma_{A,Y}(X)}{\sigma_{A,Y}} + \frac{\sigma_{X,Y}(X)}{\sigma_{X,Y}} + \frac{\sigma_{X,D}(X)}{\sigma_{X,D}} + \frac{\sigma_{A,Z}(X)}{\sigma_{A,Z}} + \frac{\sigma_{Z,Y}(X)}{\sigma_{Z,Y}}$$

$$C_B(X) = \frac{1}{1} + \frac{1}{1} + \frac{1}{1} + \frac{1}{1} + \frac{1}{1} + \frac{0}{1} = 5$$

To calculate betweenness centrality is quite expensive. The complexity of this is O (N^3)

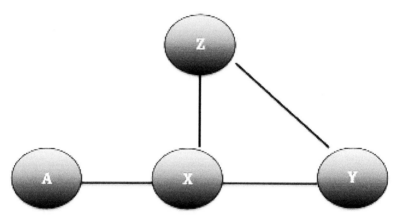

FIGURE 4.4 Betweenness centrality.

4.2.3.4 EIGENVECTOR CENTRALITY

In this measure, the importance of a node depends on how much the neighbors of that node are essential. This is used to measure the influence of an actor in the network. Centrality is proportional to the sum of all neighbor's centrality. This method can be used in the neural network (NN) (Fletcher and Wennkers, 2017). The Eigenvector centrality of the node j can be calculated as (Bonacich, 2007).

$$Ax = \lambda x$$

where λ is a constant value and $\lambda x_j = \sum_{i=1}^{n} a_{ji} x_i$; $a_{ji} = 1$, if there is a connected edge between j and i else, it is 0.

4.2.3.5 ECCENTRICITY CENTRALITY

This is the global centrality measure, which determines the node, which can circulate the information between the nodes fast (Hage and Harary, 1995). It can be calculated as:

$$C_{EC}(x) = 1 / max_{y \dot{\in} N} d_{x,y}$$

The problem in this is that it can consider only the maximum distance (Das, 2018).

4.2.3.6   INFORMATION CENTRALITY

This measure can calculate the information about the entire path between two nodes (Stephenson and Zelen, 1989). According to Stepson, the information in the path is the reciprocal of the distance. This is an advantageous measurement where the information is passing also has equal importance as the node. It can be calculated as:

$$C_{IC}(x) = \frac{1}{b_{xx} + T - 2R/n}$$

where: $T = \sum_{j}^{n} b_{jj}$ and $R = \sum_{j}^{n} b_{ij}$ for any node i.

B = $(D - A + U)^{-1}$ in this D is Diagonal Matrix with degree value and U is Unitary Matrix (Das, 2018).

4.2.3.7   STRESS CENTRALITY

This centrality can measure the amount of total amount of communication that can pass through the node or edge (Shimbel, 1953). It can be calculated as:

$$C_s(x) = \sum_{y \notin N} \sum_{z \notin N} \sigma_{yz}(x)$$

$\sigma_{yz}(x)$ Denotes the shortest path between y and z. 'x' can be a node or edge.

4.2.3.8   KATZ CENTRALITY

This method is used in the WWW (Newman, 2010). It is the general form of eigenvector centrality. It is used to calculate the influence by calculating the number of walks between two actors (Kartz, 1953). The formula to calculate is:

$$C_k(x) = \sum_{k=1}^{\infty} \sum_{j=1}^{n} \alpha^k (A^k)_{ji}$$

where A is the adjacency matrix in which the value of a_{ji} is 1 if there is a direct connection between j and i else it will be 0. The value of α will always be less than the inverse of the absolute largest Eigenvalue of A.

4.2.3.9   SUBGRAPH CENTRALITY

This centrality gives the total no of closed walk having length k starting and ending at node i. This is given by the local spectral movement $\mu_k i = A^k_{ii}$ (Estrada et al, 2005). This measure is important as when the number of close walks will increase then centrality's influence will decrease (Das, 2018).

It can be calculated as:

$$C_{SG}(i) = \sum_{k=0}^{\infty} \frac{\mu_k(i)}{k!}$$

The generalized form of subgraph centrality is also called functional centrality; mathematically it can be calculated as (Rodriguez et al., 2006):

$$C_f(i) = \sum_{j=0}^{\infty} a_j \mu_j(i)$$

The no of the closed walk of length l can be calculated by $a_l * \mu_k$.

4.2.3.10   K-SHELL CENTRALITY

In this, the whole web network is divided into K-shell structure. In it, we remove all the nodes whose degree is k = 1 and assign k_s = 1. Similarly, we remove all the nodes of degree k = 2 and assign k_s = 2 and so on (Das, 2018). This process will repeat again and again till all the nodes in the network all not considered. K shell centrality will be measured by k value. This method is important to find out the influential node that can take the responsibility of spreading the process (Liu et al., 2014).

4.2.3.11   NEIGHBORHOOD CORENESS CENTRALITY

This centrality considers that the node having more links to the neighbors located in the core of the web network will be more critical (Bae and Kim, 2014). The network core centrality can be measured mathematically:

$$C_{NC}(x) = \sum_{w \in N(x)} ks(w)$$

where $N(x)$ is the set of neighbors of node x and $ks(w)$ is the neighbor node w.

4.2.4 PAGERANK

This is very useful in measuring the significance of WebPages in the web network. The score of page rank depends on the number of in-links of that page. The page rank a node will depend on the page rank of the other nodes (Page et al., 1998). The process of calculating the page rank is:

1. Assign all the nodes page rank $1/N$ initially where N is the total number of nodes.
2. Update the page rank k times with a rule that each node will give an equal share of its page rank to all the nodes it links to. The new page rank will be the sum of all the page ranks it receives from other nodes.

The node having the highest page rank will be considered as the most critical node in the network. The page rank can be more explained with the following example:

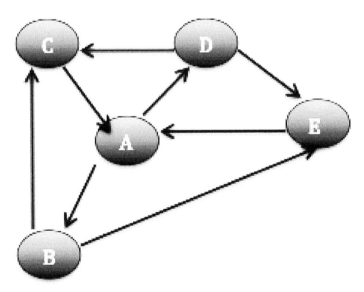

FIGURE 4.5   Pagerank.

K	Page Rank				
	A	**B**	**C**	**D**	**E**
1(initial)	1/5	1/5	1/5	1/5	1/5
2	2/5	1/5	1/6	1/10	1/5

- **Step-1:** Initially, all the nodes will get the equal page rank it means all nodes are equally important.
- **Step-2:** Then we again calculate the new page rank depending on the previous page rank and considering the equal share of page rank given to all the nodes it links to. The process can be repeated fork = 2, 3, 4…so on.

$$FOR\ K = 2$$
$$Page\ Rank\ of\ A = 1/5 + 1/5 = 2/5$$
$$Page\ Rank\ of\ B = 1/5$$
$$Page\ Rank\ of\ C = (1/2*1/5) + (1/3*1/5) = 1/6$$
$$Page\ Rank\ of\ D = (1/2*1/5) = 1/10$$
$$Page\ Rank\ of\ E = (1/2*1/5) + (1/2*1/5) = 1/5$$

4.3 MACHINE LEARNING (ML)

The techniques used in ML are used to explore the unexpected or hidden patterns in training data. ML not only works on unpredicted and concealed patterns but also to perceive the intended meaning of the processes that produce the data (Chen, 2011)

ML can be categorized in mainly three types: supervised learning, unsupervised learning, and reinforcement learning (RL). By category, there are different types of techniques are used like classification, regression, clustering, etc. (Figure 4.6).

4.3.1 LEARNING PARADIGMS

In ML, supervised learning, unsupervised learning, and RL are the learning paradigms. These paradigms influence the training data.

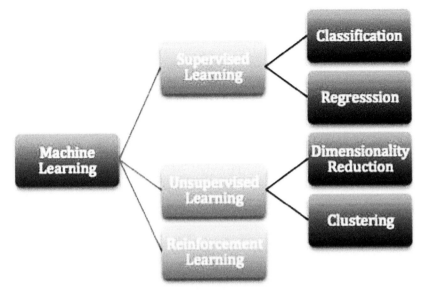

FIGURE 4.6   Learning paradigms of ML.

4.3.1.1   SUPERVISED MACHINE LEARNING (ML)

In supervised ML, algorithms are created that produce general hypotheses by inputs (Kotsiantis, 2007). Each dataset instance of ML algorithm is a representation of the same set of instances. If the instances are given with corresponding correct output, then that is called supervised (Kotsiantis, 2007). In SML, if x is the given input and y is the output, then an algorithm will be generated to find out how the mapping from input to the output can be done.

$$Y = f(x)$$

The process that can be followed while doing supervised ML is given in Figure 4.7.

In supervised ML, first, check that what is the actual problem and what are the objectives of the problem. Then after knowing the problem, the next step is to identify the most related data from different sources. The sources can be anything like social sites, search engines, international bodies and agencies, government, etc. After collecting the relevant data one of the critical tasks comes, that is, Data Preprocessing. There are a number of the task comes in data preprocessing like the integration of data (convert the data in the same format), Removal of the duplication of data (as the same data can be generated from the different sources if any take only one copy), removal of noisy data, etc. Then based on the problem, training data will be selected then SML will decide the algorithm based on the data and provided will be tested on the data. If the data are tested if it is according to the user's expectations, then the classifier will be made else it will move to the performance tuning phase or any other previous phase according to the type of problem.

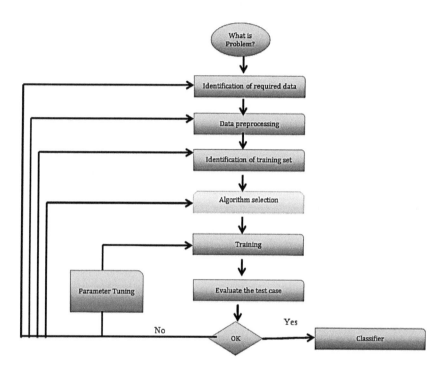

FIGURE 4.7   Supervised machine learning (Akinsola, 2017).

SML is mainly used in classification and regression. It is the process of prediction of unknown output values by using some known input values with the help of the model (Kotsiantis, 2007). For this purpose, various types of algorithms can be used like a decision tree (DT), Naïve Bayes classifier, SVM, deep learning, linear regression, etc. Which algorithm will be applied that will purely depend on the nature of the task, e.g., if work is on multi-dimensions and continuous features then SVM and NN can be used. For some applications like speech recognition and detection, face recognition, etc. deep learning is best (Kotsiantis, 2007).

4.3.1.2 UNSUPERVISED LEARNING

In this type of ML, the unlabeled data will act as input and with the help of different algorithms; unknown patterns or structures are discovered. This learning technique does not contain pre-labeled data. The techniques that can be used in this are clustering, association, and dimensionality reduction. The commonly used algorithm that can be used Apriori and K-means (Kosala et al., 2000).

4.3.1.3 REINFORCEMENT LEARNING (RL)

It enables learning through feedback also called reinforcement signal received by interacting with the outer environment. In this learning technique, the machine can automatically determine the ideal behavior within a specific context based on that feedback. DT and NN techniques can come under this category (Kosala et al., 2000).

4.3.2 COMPONENTS OF MACHINE LEARNING (ML) SOLUTION FOR NETWORKING

The components of ML to design the solution of the web network problem are discussed in subsections (Figure 4.8).

4.3.2.1 DATA COLLECTION

The learning paradigm has a significant influence on data collection and feature engineering. There is a strong relationship between learning paradigm, problems, and training data (Boutaba et al., 2018). The data collection plays an important role to handle the web network problem as this data can vary not only from problem to problem but also from time to time. In handling the web network problem, the online data collection can be used to retrain the ML model and also can be used as the feedback for the model. In the case of active network monitoring, the network structure can collect data.

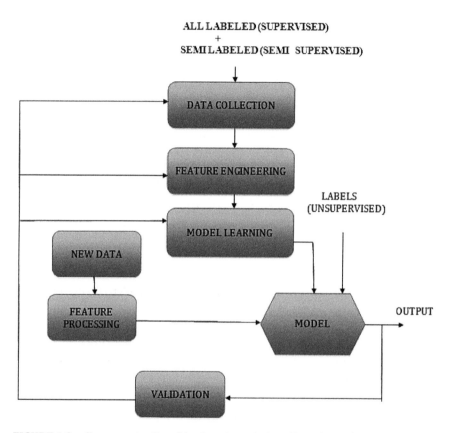

FIGURE 4.8   Components of machine learning solutions (Boutaba et al., 2018).

4.3.2.2   *FEATURE ENGINEERING*

Feature engineering includes feature selection and extraction. It can be used to reduce the dimensionality of collected web data, which can result in a reduction of computational overhead and leads to a more accurate ML model.

4.3.2.3 MODEL LEARNING

To establish an ML model, it is essential to understand the real-time scenarios. Experts must have to understand the application signature of that model before designing it, e.g., to design the ML model for handling the web networking various application signatures like the most influential actor, the flow of data between the two actors, the density of web network should be considered. There are some web network measures like centrality; CC can help here to make an accurate decision (Sagar et al., 2012).

4.3.2.4 VALIDATIONS AND PERFORMANCE METRICS

Once the model has been made, the next step is to validate the model. There are many performance matrices like accuracy, robustness, complexity, and reliability, which can be used to check the efficiency of the model. In networking, problem accuracy is a critical aspect. Accuracy can be used to improve the robustness of the model (Sagar et al., 2012).

4.4 MACHINE LEARNING (ML) FOR WEB NETWORK ANALYTICS

Web network analysis includes analysis of websites, web servers, web traffics, page views, web transactions, etc. It is used to deeply understand the usage of the web and also in helping to optimize system performance (DINUCĂ, 2011). ML provides a novel insight in Web network by text classification, text clustering, relevance feedback, information extraction, etc. Various ML algorithms can be used to analyze web data. ML is best suited for analyzing the structured, unstructured, and semi-structured data. There are various ML classifiers, which are used to analysis multimedia data like audio, video, etc. The accuracy of ML depends on how good training data is. By inspecting the inter-links and intra-links in the web network, ML model can come to know about the expected results. Depending on hyperlink topology, with the help of ML techniques, the websites can be proficiently categorizing that which is relevant or not.

The primary goal of ML in the web network is to produce the summary of web sites and web pages. Based on the training data ML can generate similar types of websites and how they are related to each other. One of the other advantages of ML is in discovering the web document structure, which can produce fruitful results in integrating, comparing, and navigating between the web pages. ML models measure the frequency of interior (links within the same document) and exterior (links with other websites) links of web tuples in web table, which can help in deciding the central actor of the web network, and also it can decide about the neighbors in Web network. The related actors may contain similar data and can be connected to the same web server. It can help in bringing together the same nature type actor and processing the query on the web more easily and efficiently. There are many measures in the Web network analysis like degree centrality, betweenness centrality, Closeness Centrality, which can be used while designing the ML model for traffic congestion. If there is, a chance of congestion Model can automatically choose the other way. This will be a great help in handling the load in the web network.

ML is actively helpful in analyzing web networking. ML can play an essential role in:

A) To discover the node that prevents the network from breaking.
B) To find out the relevant pages on the Web network.
C) To find out which actor can act as hub currently.
D) To find out the node that spread information to many other nodes.
E) To find out the node that can prevent epidemics.
F) To find out the most critical node in the web network.

4.5 CONCLUSION

The web is the world's biggest knowledge repository and extracting related data from the Web is an enormous challenge. Many companies are using ML to deal with the web as machine-learning can not only detect the patterns from the vast amount of data but also can automatically change the action on the base of the detected pattern. ML is very helpful in understanding the customer behavior and it can also deliver the personalized content and information to the user by detecting patterns by user's likes and posts (Chen, 2011). In this chapter, ML is providing a close insight into the web network by finding the most important actors, clustering, and path finding, etc. For this in ML, various networking analytics measures like Page Rank, centrality, CC can be used which are themselves can offer a proper assessment on relationships and information flow and very useful in optimizing the nodes.

KEYWORDS

- **betweenness**
- **clustering coefficient**
- **feature engineering**
- **model learning**
- **performance metrics**
- **web servers**

REFERENCES

Adamic, L. A., & Adar, E., (2003). Friends and neighbors on the web. *Social Network, 25,* 211–230.

Ahmad, M., Aftab, S., Muhammad, S. S., & Ahmad, S., (2017). Machine learning techniques for sentimental analysis: A review. *International Journal of Multidisciplinary Sciences and Engineering, 8*(3), 27–32.

Akinsola, J E T., (2017). *Supervised Machine Learning Algorithms: Classification and Comparison, 48*(3), 128–138.

Bader, D. A., Kintali, S., Madduri, K., & Mihali, M., (2007). Approximating betweenness centrality. *WAW'07 Proceedings of the 5th Workshop on Algorithms and Models for Web-Graph* (pp. 124–137). Springer-Verlag Berlin, Heidelberg.

Baes, J., & Kim, S., (2014). Identifying and ranking influential spreaders in complex networks by neighborhood coreness. *Physica A., 395,* 549–555.

Bansal, H., Shrivastava, G., Nguyen, G. N., & Stanciu, L. M., (2018). *Social Network Analytics for Contemporary Business Organizations*. IGI Global.

Benchettara, N., Kanawati, R., & Rouveirol, C., (2010). Supervised machine learning applied to link prediction in bipartite social networks. *International Conference on Advance Social Network Analysis and Mining, IEEE,* pp. 326–330.

Bonacich, P., (1987). Some unique properties of eigenvector centrality. *Social Networking, 29,* 555–564.

Boutaba, R., Salahuddin, M. S., Limam, N., Ayubi, S., Shahriar, N., Solano, F. E., & Caicedo, M., (2018). A comprehensive survey on machine learning for networking: Evolution, applications and research opportunities. *Journal of Internet Services and Applications, 9*(16), 1–99.

Brandes, U., & Pinch, C., (2007). Centrality estimation in large networks. *J. Bifurcation and Chaos in Applied Science and Engineering, 7,* 2303–2318.

Chen, H., & Chau, M., (2005). *Web Mining Machine Learning for Web Applications, 38*(1), 289–329.

Cooley, R., Mobasher, B., & Srivastava, J., (1997). Web mining: Information and pattern discovery on the world wide web. *Proceedings of International Conference on Tools with Artificial Intelligence*, pp. 558–567.

Das, K., Somanta, S., & Pal, M., (2018). Study on the centrality measure on the social network: A survey. *Social Network Analysis and Mining, 8*(1), 1–11.

Dinucă, C. E., (2011). Web structure mining. *Annals of the University of Petroşani, Economics, 11*(4), 73–84.

Er Hari, K. C., (2017). Online social network analysis using machine learning techniques. *International Journal of Advances in Engineering and Scientific Research, 4*(4), 25–40.

Estrada, E., & Rodriguez-Velazquez, J. A., (2005). *Subgraph Centrality in Complex Network, 71*, 1–29.

Fletcher, J. M., & Wennekers, T., (2017). From structure to activity: Using centrality measures to predict neuronal activity. *International Journal of Neural System, 27*, 1750013–1750016.

Freeman, L. C., (1978). *Centrality in Social Networks: Conceptual Clarification* (Vol. 1, No. 3, pp. 215–239). Social Network, Elsevier Sequoia.

Hage, P., & Harary, F., (1995). Eccentricity and centrality in network. *Social Networking, 17*, 57–63.

Hastie, T., Tibshirani, R., & Friedman, J., (2011). *The Elements of Statistical Learning: Data Mining, Inference, and Prediction.* Springer, New York.

Jordon M. I., & Mitchell, T. M., (2015). *Machine Learning: Trends, Perspective and Prospects, 349*, 255–260.

Kartz, L., (1953). A new status index derived from sociometric analysis. *Psychometrika, 18*(1), pp. 39–43.

Kosala, R., & Blockeel, H., (2000). *Web Mining Research: A Survey* (pp. 1–15).

Kotsiantis, S. B., (2007). Supervised machine learning: A review of classification techniques. *Informatica, 31*, 249–268.

Kumar, D. G., & Gosul, M., (2011). In: Wyld, D. C., et al., (eds.), *Web Mining Research and Future Directions* (Vol. 196, pp. 489–496).

Kumari, R., & Sharma, K., (2018). Cross-layer based intrusion detection and prevention for the network. In: *Handbook of Research on Network Forensics and Analysis Techniques* (pp. 38–56). IGI Global.

Liu, J. G., Ren, Z. M., & Guo, Q., (2014). Ranking the spreading influence in complex networks. *Phys. A., 392*(18), 4154–4159.

Murphy, K., (2012). *Machine Learning: A Probabilistic Perspective.* MIT Press, Cambridge, MA.

Newman, M. E. J., (2010). *Networks: An Introduction.* Oxford University Press, New York.

Page, L., Brin, S., Motwani, R., & Winigrad, T., (1998). The page rank citation ranking. In *Bringing order to the Web* (pp. 25–49). Technical report, Springer Standford University, Stanford, London.

Plazzl, R., Baggio, R., Neldhardt, J., & Werthne, H., (2012). Destinations and the web: A network analysis view. *Information Technology and Tourism*, 215–228.

Rodriquez, J. A., Estrada, E., & Gutierrez, A., (2006). *Functional Centrality in Graph* (Vol. 55, pp. 293–302). Linear Multilinear Algebra.

Sabidussi, G., (1966). The centrality index of a graph. *Psychometrika, 31*(4), 581–603.

Sagar, S., De Dehuri, S., & Wang, G. N., (2012). Machine learning for social network analysis: A systematic literature review." *The IUP Journal of Information Technology, 7*(4), 30–51.

Sansanwal, K., Shrivastava, G., Anand, R., & Sharma, K., (2019). Big data analysis and compression for in air quality. *Handbook of IoT and Big Data*, 1.

Sharma, K., & Shrivastava, G., (2014). Public key infrastructure and trust of web based knowledge discovery. *Int. J. Eng., Sci. Manage.*, *4*(1), 56–60.

Shimbel, A., (1953). Structural parameters of communication networks. *Bull. Math. Biophys.*, *15*(4), 501–507.

Shrivastava, G., & Bhatnagar, V., (2011a). Analyses of algorithms and complexity for secure association rule mining of distributed level hierarchy in web. *International Journal of Advanced Research in Computer Science*, *2*(4).

Shrivastava, G., & Bhatnagar, V., (2011b). Secure association rule mining for distributed level hierarchy in Web. *International Journal on Computer Science and Engineering*, *3*(6), 2240–2244.

Shrivastava, G., (2017). Approaches of network forensic model for investigation. *International Journal of Forensic Engineering*, *3*(3), 195–215.

Shrivastava, G., Sharma, K., & Bawankan, A., (2012). A new framework semantic web technology-based e-learning. In: *2012 11ᵗʰ International Conference on Environment and Electrical Engineering IEEE*, 1017–1021.

Wadhwa, P., & Bhatia, M. P. S., (2012). *Social Networks Analysis: Trends, Techniques, and Future Prospects, Fourth International Conference in Recent Technologies in Communication and Computing*, 1–6.

Zhao, Z., & Liu, H., (2007). Spectral feature selection for supervised and unsupervised learning, ICML'07. *Proceedings of 24ᵗʰ International Conference on Machine Learning*, 1151–1157.

CHAPTER 5

Online Trust Evaluation

DEEPAK KUMAR SHARMA, NIKITA CHAWLA, and
RAVIN KAUR ANAND

*Division of Information Technology, Netaji Subhas Institute of
Technology, New Delhi, India, E-mails: dk.sharma1982@yahoo.com
(D. K. Sharma), nikitachawla3@gmail.com (N. Chawla),
ravinkaur97@gmail.com (R. K. Anand)*

ABSTRACT

In the recent years, we have also started witnessing the rise of a novel paradigm-the Internet of Things (IoT), which describes the idea of everyday physical objects being connected to the internet and being able to identify themselves with other devices. While this fosters the idea of "ambient intelligence", it also introduces new concerns for risks, privacy and security, owing to which trust evaluation becomes imperative in this environment. As the backbone of IoT is the ability to make decisions without human intervention, trust serves as a crucial factor that can support the establishment of reliable systems and secure connectivity. It could help both the infrastructures and services in managing risks by operating in a controlled manner and avoiding unpredicted conditions and service failures. This chapter deals with defining trust and its characteristics, mainly in an online environment and further explores the role of trust evaluation in areas mentioned above such as e-Commerce, Online Social Networks (OSNs) and the Social Internet of Things (SIoT). It looks at the various algorithms, mechanisms and models proposed for trust evaluation in these domains and identifies the major roadblocks that may be encountered during their implementation. Towards the end, it discusses how digitalization, on one hand, may have connected the world, made processes easier, it has also, on the other hand, increased susceptibility to fraud, and security breaches making it even more important to overcome challenges to establishing trust online.

5.1   AN OVERVIEW OF ONLINE TRUST EVALUATION

The fundamental basis of all human interactions and social activities has always been trusted. Even in business relationships, no matter what industry the business belongs to, trust has always played a key role. Without the notion of trust, our day to day activities may be prone to high risk and deceit, making it necessary for us to ponder upon how a trust may be measured and evaluated comprehensively.

In our lives today, the use of technology has increased more than ever before. To make our life smoother and faster, we resort to using various online platforms for fulfilling our personal and business goals. Be it watching videos or listening songs to making payments, the reliance on the human race on technology has evolved in more ways than one. This brings newer dimensions to the traditional notions of trust as loyalty and satisfaction in the online environment are different from that in the offline context. There is a general agreement that it is more challenging to build trust online than offline; however, once built, it can lead to improved efficiency, greater flexibility and reduced costs for the firm.

Hence, the concept of trust has evolved more than ever in the last decade and people have started becoming aware of the uniqueness of trust in an online environment, redefining it as *online trust.* With the advent of social networking, e-commerce, and new-age technologies like the Internet of things, cloud, and big data (BD), the question on how a trust may be efficiently evaluated online, however, remains an ongoing area of research. Studies have been conducted on how, with the increase in the use of Social Networking Sites as a tool to overcome geographical barriers and to make friends, graphs called online social networks (OSNs) have are being used increasingly to evaluate the trustworthiness of the users. Similarly, with the boom in the e-commerce industry with the advent of start-ups like Flipkart, there is also a concern about the trust between the buyers and sellers, which is addressed using methods of online trust evaluation.

Against the above-mentioned background, this chapter addresses three objectives: (1) to explore the concept of trust and examine the characteristics of trust when viewed in the context of an online environment; (2) to illustrate how trust can be built online and how its evaluation is different from Risk Management; and (3) to analyze the various approaches to evaluating trust online in varied digital spaces, namely e-commerce, OSNs and the social internet of things (SIoT).

5.2 WHAT IS TRUST?

Defining trust concisely is a challenging task. This is because it is unique to each entity and varies from individual to individual. In most cases, trust is defined as the potential and belief of an entity that the other entity would be able to meet its expectations. In a prominent work to define the notion of trust, a person is considered trustworthy in terms of the probability that he will perform an action that is worthwhile to us.

Trust has specific unique characteristics to it when viewed in different contexts, i.e., in an online environment and an offline environment. In an online environment, trust can be categorized into two brackets: direct trust between two entities involved in a transaction happening directly between them and third-party trust which is a relationship formed from third-party recommendations. Direct trust assumes that entity A trusts entity B after several successful transactions have occurred between them, while in a third-party relationship, there is no interaction between the involved parties A and B, and the trust between them is established with the help of entity C which is in a direct relationship with both of them or in a third-party relationship with a directly trusting entity.

5.3 CHARACTERISTICS OF ONLINE TRUST

Online and offline trust have various similarities in terms of their characteristics, but there are certain distinctions which make the online environment unique. The characteristics of online trust and their comparison with that of offline trust are as follows:

5.3.1 TRUST OR AND TRUSTEE

The two entities trustor (a party that trusts) and trustee (a party that is trusted) play a very crucial role in the establishment of a trusting relationship in an online as well as the offline environment. In the offline context, different entities such as people, organizations, or even products can play the role of the trustor and/or trustee depending on the given situation. However, in an online backdrop, usually, the trustor is a consumer who is browsing websites looking for products and services while the trustee is a website (merchant) that provides the consumer with products and services. A consumer (trustor) will buy a product if only if he believes in the trustworthiness of the merchant (trustee). A trustful relationship is thus dependent on how the best interests of the trustor can be fulfilled by the trustee and to what degree.

5.3.2   *VULNERABILITY*

Trust is a property which only flourishes when there is an element of vulnerability and uncertainty in an interaction. Trustors must have an appetite for taking the risk that may result in losing information crucial to them so that they become vulnerable and should rely on the trustees that their vulnerability will not be exploited. The highly complex nature and anonymity associated with trust on the Internet can also cause the trustee to behave maliciously. Therefore, many times consumers (trustors) are uncertain about the risks of transacting/interacting online. Sometimes even when users are just surfing the online environment, they are susceptible to threats that involve data about their activities to be automatically collected and misused without their knowledge. These reasons thereby become a basis for an insecure online environment in which trust evaluation is required and can thrive.

5.3.3   *PRODUCED ACTIONS*

Trusting an entity always leads to action and usually these actions involve taking the risk. The type of action is situation-dependent and may involve exchanging something that may be tangible or intangible. For example, someone can borrow money from his friend or relative because they trust him to return his money. Similarly, when operating online, consumer trust in merchants comes into action. This trust can bring about two forms of action from the consumer which is:

    i.   Sharing credit card details while making a purchase online along with other personal information; and

   ii.   Simply just surfing the website.

To perform these activities, the consumer needs to have trust in the intent of the merchant and on the fact that the probability of gaining is more than getting harmed.

5.3.4   SUBJECTIVE MATTER

Trust has an inherent element of subjectivity. It is severely affected by small changes in individual preferences and situational factors. Different people have a different concepts of trust and might view trust differently in varying scenarios. They may also have different intensities of trust when different trustees are considered. Similar to the offline environment, the level of trust online also varies from individual to individual. This is because people also have different opinions and attitudes toward various technologies and machines.

5.4   WEB INTERFACE DESIGN DIMENSIONS FOR ONLINE TRUST

When it comes to building the trust of consumers online, one of the most effective ways that online merchants should consider is having a well-thought web interface design. It is the most efficient way of turning first-time visitors on their websites into loyal and profitable customers as the first thing that a customer uses to judge an entity online is the user-interface of its website or application. Since there is no appearance of buildings, employee engagements in the online business context, online merchants need to rely on a variety of trust-inducing features in their web interface design. There are four broad dimensions in which trust-inducing features for establishing trust online can be categorized into namely, graphic design, structure design, content design, and social-cue design. These dimensions are further discussed below.

5.4.1   GRAPHIC DESIGN

It refers to the factors related to graphic design on the website. As graphics are the first visible elements on opening a website or application, they are the ones that make the first impression. Sizes of the clipart, well-chosen photographs, and color-schemes, all create an impact on the visitor and play a key role in developing his sense of confidence in the entity.

5.4.2   STRUCTURE DESIGN

It refers to the accessibility and overall structure of the information displayed on the website. The more easily accessible and well-displayed the information is, the easier it is for the user to navigate through the website or application and find essential information, which further makes it seem more trustworthy to the user. Simplicity, consistency, and a set of tutorials or instructions for navigation and accessibility remain crucial features of good interface design.

5.4.3   CONTENT DESIGN

This is one of the most critical aspects of web-interface design when viewed from the trust perspective. It refers to the content displayed on the website, both textual and graphical. Elements like third-party certificates, approval seals, etc. play an essential role in trust establishment of the customer. Similarly, clear, and lucid display of customer relationship information, including all disclosures relating to financial, privacy or legal concerns, is also effective.

5.4.4   SOCIAL-CUE DESIGN

This dimension involves eliminating the lack of "human touch" or online presence by including social-cues like presence on social media and face to face interaction options into the application design through different modes of communication. The interactivity of communication media, functionalities of the embedded social-cues like chat, call-back opportunities, representative photographs, or video clips all add to increasing the general perceived level of trust among the consumers.

5.5   TRUST EVALUATION VERSUS RISK MANAGEMENT

There is a common misconception that trust evaluation and risk management refer to the same thing, which is assessing the degree of confidence in completing a goal. Indeed, trust evaluation and risk management have a strong correlation. However, both terms encompass different factors within them. When we talk about managing risk in the digital space, we refer to the complex process of investigating vulnerabilities, threats, and risks in physical networking systems. This process includes framing the risk, assessing it, responding to it once determined and monitoring it.

Although all these steps are also common to the trust evaluation process, they are seen and performed from a different perspective than that of risk management. Trust Evaluation is more human-oriented; thus, takes into account a lot of social factors like opinions, experiences, etc. of the participants. Risk Management, on the other hand, is more system-oriented and assesses only the computer system as an entity. Trust considers the entity from under the lens of the trustor and is used to express a very subjective opinion about the trustee in a specific social context, while in the case of Risk Management; the context is only the digital system (see Figure 5.1).

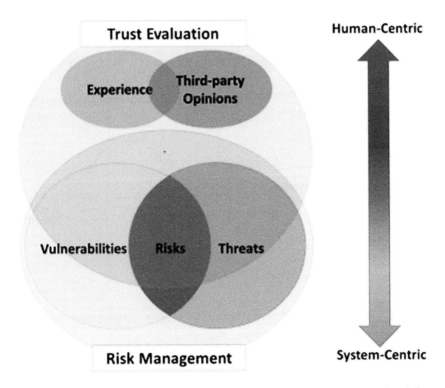

FIGURE 5.1 Comparison of risk management and trust evaluation. (Reprinted from Truong et al., 2017. http://creativecommons.org/licenses/by/4.0/).

5.6   ONLINE TRUST EVALUATION IN E-COMMERCE

5.6.1   DEFINING E-COMMERCE AND IMPLICATIONS OF TRUST?

E-commerce is a medium through which consumers and merchants may purchase or sell products and/or services online, i.e., over the Internet, using digital communication as an option for payment (see Figure 5.2.) Since the development of the internet, several models have been identified to define the different types of e-commerce such as B2B (business to business), B2C (business to consumer), C2B (consumer to business), C2C (consumer to consumer), B2G (business to government), and Mobile Commerce. Over the years, B2C e-commerce has seen phenomenal growth as many organizations have shown growing interest to use Electronic Commerce to enhance their competitiveness and cater to a broader audience.

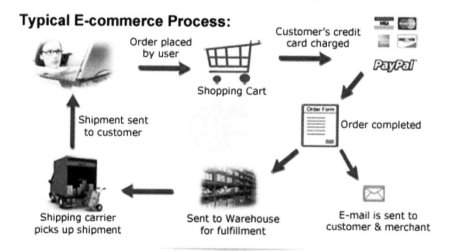

FIGURE 5.2   Depiction of a typical e-commerce process. (Reprinted from https://www.pinterest.com/pin/322359285822573233)

E-commerce has gained wide popularity because of the benefits it offers in contrast to traditional commerce. These benefits may be attributed to the availability of more quantity and quality of information to the customers. For Example, they may gather as much information required via user-friendly websites, track their orders, and/or provide feedback.

However, the Internet's open system environment along with the innovation that comes with remote shopping is also prone to some hindrances in terms of generating consumer loyalty, the most significant being lack of trust.

The concept of trust is thus, is given much importance by researchers as if there is no general notion of online trust, doing business online would be a challenge. The various threats that exist in the e-commerce space can be seen from two perspectives: business-related, and technology-related (see Figure 5.3). Technology-related risks include security, integrity, and privacy concerns, while on the other hand; the business-related risks include those that involve the misuse of personal information and fraudulent transactions. E-commerce systems tend to be unstable because of such risks. Consumers, therefore, become apprehensive of the low level of personal data security, hidden charges, and difficulties in refunds/cancellations and of the possibilities of non-delivery of goods. With these risks and challenges prevalent in the online sphere, all the online merchants are majorly dependent on their web presence via websites and applications only to represent themselves to prospective clients. This makes them focus on web interface extremely integral to online selling. It also calls for the use of different frameworks like reputation-based algorithms and feedback analysis by the sellers to enhance their perceived trustworthiness to potential customers. All major e-commerce websites today like Amazon, E-bay are actively integrating these models within their systems to evaluate customer trust and enhance their processes, accordingly, thereby making this a popular field of study within their research teams.

5.6.2 REPUTATION BASED ONLINE TRUST EVALUATION

In applications involving e-commerce, the reputation of the seller is a major concern for customers while making a purchase or payment. This has made researchers extremely interested in reputation-based search algorithms. Trust evaluation using reputation-based schemes usually focuses on parameters like service type, transaction, and interaction history. Socially oriented environments demand that the focus while evaluating trust should lie on the nature of the relationship between the interacting parties while service-oriented environments require that there is reasoning and maintenance about quality of services (QoS) in addition to relationships when evaluating trust. The latter also requires that recommendation-based trust is evaluated and incorporated in the context.

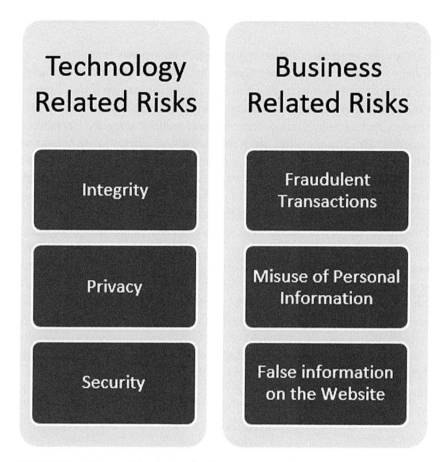

FIGURE 5.3   Various risks associated with e-commerce.

Reputation-based trust evaluation can be applied not only to socially-oriented trust environments but also to service-oriented trust environments. This can be understood by observing how a service can gain a good reputation. This is usually done via customer ratings, once services of good quality are accumulated over some time; ratings improve thereby signaling a good reputation. However, this does not solely constitute the final reputation value. To come to a final reputation value, the relationship between the trustee (who gets rated) and the trustor (who rates) is studied, and more objective trust results are obtained.

We further discuss the reputation-based analysis in different trust-computing categories.

5.6.2.1 REPUTATION EVALUATION IN C2C E-COMMERCE

To study the reputation evaluation in C2C e-commerce, we consider one of the earliest yet best existing algorithms, i.e., that of e-commerce giant eBay. In eBay's trust management mechanism, customer feedback is given the utmost importance. After every purchase, the customer can rate the service quality as one of the three available options of positive, neutral, or negative. This rating is then stored along with all other customers' ratings and used to calculate a feedback score for the firm. The feedback score calculation is simply the difference between the number of positive ratings and Negative Ratings left by the customers. This score is then published and updated regularly on the web page of the seller, acting as plain data to the buyers for coming to a judgment on their own. Similarly, it uses encouragement metrics like a positive feedback rate to reward sellers with a positive feedback rate greater than a specified threshold, also referring to them as power sellers. In this manner, we can see that this reputation-based scheme is fairly simple yet very effective and gives the users a fair chance to know about sellers and their credibility.

5.6.2.2 TRUST EVALUATION IN PEER TO PEER NETWORKS

In P2P information-sharing networks, before downloading data, the client peer should have the relevant information to know which service peer can fulfill its needs for files. For this, the method of trust evaluation may use polling algorithms, a system for calculating final trust value by binary rating or a voting reputation mechanism that requests the experience of other peers with the given peer and combines the values given by the responders to come to a final value of trust. In the case of systems involving file sharing, binary-value ratings are used to signify if the file is a complete version or not. For more complex applications, like the ones that are service-oriented a rating in the range of [0,1] is used.

5.6.2.3 TRUST EVALUATION IN MULTIAGENT ENVIRONMENTS

When we talk about evaluating trust in multi-agent environments, i.e., in cases where autonomous and self-interested software agents finish the tasks given by their owner or even other agents, other issues crop up in online trust evaluation. These include evaluating the agent's motivation and also the dependency relationships between them. Various models like the multidimensional trust model, generic trust vector models, and others that may be based on the rank and reputation of QoS-based services are used in multiagent collaboration situations.

With the above-given models and mechanisms for evaluating trust online in e-commerce related services and transactions, the user may use a suitable architecture based on factors of choice like workload, scalability, reliability, and nature of tasks.

5.6.3   *OLX.NG: A CASE STUDY ON TRUST IN THE E-MARKETPLACE*

5.6.3.1   *OLX.com: THE WORLD'S ONLINE MARKETPLACE*

With the boom in the e-commerce industry, several companies started to come up since the early 2000s to act as e-marketplaces, with OLX being a major disruptor. Founded in 2006, OLX.com is an online platform for users to buy, sell, and exchange a variety of products. From advertising, it has come up with novel ways through which users can easily design catchy advertisements and use the concepts of control buying-selling and community engagement through the addition of pictures, videos, and display options on social networking sites like Facebook, LinkedIn, etc. Spread across over 100 countries and available in more than 30 languages, OLX.com has grown to be of the world's largest online marketplaces today.

5.6.3.2   *A SURVEY ON TRUST BETWEEN BUYERS AND SELLERS*

5.6.3.2.1   *The Survey*

A group of final year students of an E-Business Course at the University of Ilorin, Nigeria conducted a survey to investigate the degree of online trust in e-commerce. Each student of the class posted items to sell on OLX. ng such as handbags, wristwatches, shoes, laptops, mobile phones, and toasters. The relationship between the students, i.e., the sellers and the buyers were then closely monitored.

5.6.3.2.2 Findings

Within a week, the sellers were able to identify at least two or more prospective buyers. However, by the end of the second week, only about 11% of students were able to sell their products. This also depended upon the kind of goods being sold, with electronics being more comfortable to sell as compared to consumer goods like handbags and shoes, showing that it is difficult to generate trust online in these cases, where there is a significant dependency on the quality, look, and feel. Similarly, about 68% of sellers received two or more calls for their products, but more than half of the buyers still did not trust the seller. One reason for this was the mode of payment provided. If the seller would not provide Cash on Delivery option, the e-vendor could be judged as one having the wrong intentions or being untrustworthy. On the other hand, this also poses a risk to the seller because there is no guarantee of payment on delivery. Therefore, there needs to be a mechanism of generating this trust online within the buyer-seller community. For example, if a seller wants to prove his trustworthiness, he may provide full information regarding his cost of goods, delivery-payment arrangements, after-sales agreements, and also his complete proof of identification on the website. Similarly, if there is a duly signed contract regarding the completion of payment, the seller's trust in the buyer would also increase. This can be achieved using digital signatures and authentication mechanisms. Hence, we see that a critical factor in the success of an e-marketplace is generating trust online. If the sellers can guarantee timely delivery of goods in good condition and the buyers can guarantee their payment at the time of the online transaction, the online trust environment can prosper.

5.7 TRUST EVALUATION IN ONLINE SOCIAL NETWORKS (OSNS)

OSNs are readily becoming a trending platform where people meet and keep in touch with family and friends. It is used by people who want to find others who share common passion and interests. Nowadays, even businesses and organizations are tapping into the world of OSNs to conduct their daily professional conversations and activities. On an online social network, interactions are very complex due to their inherent graphical structure and also because they may involve conversing with strangers, some of whom might have malicious intent. This coupled with the lack of real-life face-to-face interaction and understanding amongst the users makes the concept of trust and 'trust evaluation' very essential. Trust evaluation aids in forming an opinion about a person's or product's trustworthiness which can further help in providing better user experience and functionality. Its increasing popularity explains the interests of researchers in designing new and better models for trust evaluation which can help in decision making in diverse applications. One such method that we will discuss is Graph-Based Trust Evaluation under which the OSN is depicted as a social graph.

5.7.1   GRAPH-BASED TRUST EVALUATION

As mentioned above, the Online Social Network is represented via a graph, called a social graph. Users are referred to as nodes of the graph and the relationship or interaction between them is represented by an edge connecting them. In an OSN, nodes can have direct as well as indirect interaction or contact with other nodes. This illustrates how trust can be built in both a direct or indirect manner. Trust that is built due to direct contact is known as the first-hand trust, while on the other hand trust that is built through indirect contact, i.e., through recommendations from other nodes is known as second-hand trust. Trust is quantified in this model by assigning the edges or path between nodes are a weight value between 0 and 1, where 0 depicts 'no trust' and 1 depicts 'full trust.' These values may vary from model to model.

The nodes in the graph can have one of the three roles namely a source (or trustor), a target (or trustee), or a recommender (an intermediate node on a trusted path). The source and target can be in direct contact. In this case, they have a given trust value for the edge between them. On the other hand, if they are not in direct contact, the trust value or trusted path between them can be generated by iterative recommendations. This means that the nodes that lie on the different paths connecting them can recommend different values (on the basis of the influence and personality of the user) that may be further used by being added directly or in a weighted manner (Figure 5.4).

In Figure 5.4, s, u, v, and d are the nodes representing the users in the OSN graph. u and v are in direct contact with d and thus they have an idea about the trustworthiness of d. The node s is not in direct contact with d but is directly connected with nodes u and v, so s forms it is the opinion of d, through u and v. In this manner, several sequential and parallel paths can be merged or overlapped to build a trusted graph between s and d. This is called the process of iterative recommendations.

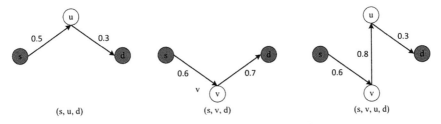

FIGURE 5.4   Examples of trusted graph.

5.7.1.1   *CATEGORIZATION OF GRAPH-BASED TRUST EVALUATION MODELS*

Trust evaluation models for OSNs are categorized by how they deal with a trusted graph. Mainly, there are two approaches: the graph simplification-based approach and the graph analogy-based approach. Models in both these approaches are discussed below:

5.7.1.1.1   *Graph Simplification-Based Approach*

Under this approach, a trusted graph is broken into multiple paths which have disjoint nodes and edges. It involves the creation of a directed series-parallel graph from a trusted graph by using the different possible paths under the guidance of some pre-defined principles. Some models that come under this category are listed below:

1. **TidalTrust:** This model uses trusted paths to produce recommendations about the degree of trust a user can put on another user. It does so by exploring trusted paths in a breadth-first search manner, beginning from the trustor and ending on the trustee. Tidal Trust uses exclusively the shortest most robust trusted paths. Trust from s to d (t_{sd}) is calculated using a threshold that has been set for being trustful and is referred to as max. The neighbor j is

considered trustful only if t_{sj} is more significant than this threshold value (max). Calculations begin from the trustor, move towards the trustee, and then the final value of the trust is returned from the trustee to the trustor.

2. **MoleTrust:** Trust evaluation through this model is divided into two steps:

   i.   Users are sorted according to their least distances from the trustor after which, cycles are detected.

   ii. Input is given which consists of a maximum depth of users from the trustor. The term maximum depth is not dependent on any specific user. Thus, the trusted graph is now a directed acyclic graph (DAG) which has been reduced.

   So according to the above steps this model initially computes the trust degree of nodes that are at a single step distance from the trustor, then two steps and similarly goes on further. So, in general, a node which is at a k-steps distance from the trustor needs only the nodes which are k-1 steps away to calculate its degree of trust. This implies that a node's details are used once. The weighted average (WA) of the trust of every incoming trustful neighbor towards the trustee is then taken to come to a final value.

3. **MeTrust:** It uses trusted paths having multi-dimensional evidence. Trust evaluation is done at three layers which are the node layer, the path layer, and the graph layer. The node layer deals with multi-dimensional trust and each node can have varying weight values for each dimension. Similarly, the second layer, i.e., path layer controls the rate at which trust decays during the combination. In the end, the graph layer simplifies trusted graphs using three algorithms: GraphReduce, GraphAdjust, and WeightedAverage. In this model, many conceivable scenarios and mutable settings can be selected flexibly.

4. **SWTrust:** This model assumes that a small trusted graph is already in existence and uses information about active domains of nodes to compute the trust value, which is much better than using explicit trust ratings. Neighbor set of a node is classified into three classes by the social distance from the trustor that are local neighbors, more extended ties, and most extended ties. Adding an adjustable width feature in breadth-first search, the next-hop neighbors from the above three given classes at each search step are chosen. This

model focuses on exploring the objective and stable information about a node such as their domain for the computation of trust and uses small trusted graphs for large OSNs.

Challenges faced by the graph simplification approach include path length limitation and evidence availability. These crop up because this approach makes use of both the composable as well as propagative nature of trust. In a trusted path, propagation means that if Node 1 trusts Node 2, and Node 2 trusts Node 3, then Node 1 can derive trust towards Node 3. Fixing an accurate path length is a challenge that the property of propagation causes, because while on the one hand, smaller path length may cause fewer paths to be discovered, on the other hand, a longer path length may lead to inaccurate prediction. The primary challenge of combining available evidence is caused due to multiple trusted paths in the graph.

5.7.1.1.2 Graph Analogy-Based Approach

In this approach, trusted graphs are emulated using other graphs and it does not involve any simplification or removal of edges or nodes from the graphs. This approach aims to scout for the similarities between graph-based models in other networking environments and graph-based trust models in OSNs, like parameters such as patterns of diffusion, network structure, etc.

Models that can be categorized under this approach are:

1. **RN-Trust:** This model uses the similarities between electric flow and trust propagation. It is known that more current can pass through the path if the resistance is lower in that path. Therefore, a resistive network is made from a trusted graph using the equation.

$$r = - \log t$$

where r refers to the resistance and t refers to the degree of trust. In order to calculate the value of trust derived finally from the trustor to trustee, the model firstly calculates the value of equivalent resistance, $R_{eq\,sd}$. Consequently, the final value of the trust is computed by using:

$$t_{sd} = 10R_{eq\,sd}.$$

2. **Appleseed:** This model borrows the idea of spreading activation models. It proposes that at the beginning, an initial amount of trust (energy in-case of spreading activation models) is injected into the source node that is the trustor and then this trust propagates into its successor nodes. The more amount of trust or energy a node gains the more trustful it gets. For handling trust decay, the model makes use of a global spreading factor δ along the trusted paths. To make the final result or value more reasonable, a normalization function is applied on the edge weight. The total truth or energy a node can keep depends upon the total energy received and a factor of global decay. While calculating energy, the model uses the backpropagation of trust from other nodes to the source node. Virtual edges are also created form every other node to the trustor or source node for this purpose. To make sure that variations in trust values are not greater than the threshold, the algorithm terminates after a fixed number of rounds.

3. **Advogato:** This model identifies the bad or unreliable nodes from the network and cuts them and the nodes that certify them out from the network. It uses the concept of trusted seeds for the computation of trust. Before any calculation, some nodes are assumed to be trusted seeds. The model then conducts a breadth-first search traversal on the OSN graph and appoints every node with a capacity, which depends on the shortest distance of the seed from the node. A seed's capacity is given as an input which represents the number of trustful nodes the system wants to search for. So accordingly, the nodes that are closer to seed will have more capacity in comparison to the nodes which are further away. Likewise, nodes that have the same distance level as the seed would have equal capacities. After the assignment of capacities to the nodes, Advogato makes use of a transform algorithm that converts a single-source/multiple-sink graph into a single-source/single-sink graph. For the above conversion to happen, a super sink is created. It divides each node into two. For example, Node N is converted into N^- and $N+$. The capacity of the edge from N^- to the created super sink is assigned a value of 1 and the capacity left of the node N is given to the edge from N^- to N^+. The capacity of an edge from N^+ to R^- i.e., to a different node, is assigned the value ∞. Lastly, the maximum integer network flow algorithm is conducted to pick the nodes which are trustful.

4. **GFTrust:** This trust evaluation scheme uses generalized trust flow where the flow is said to leak as it propagates in the network. In quantitative terms, this leakage is referred to as the gain factor of a particular edge. For example, let us consider that a flow of 2 enters an edge, and the gain factor of the edge is 0.9. Then, 2 * 0.9 = 1.8 flow will leave the edge. This model deals with overcoming the path dependence challenges using flow in the network, and the concept of trust decay due to propagation using the leakage linked with each node. For constructing a generalized network, the intermediate nodes are divided into N^+ and N^-, an intermediate edge between N^+ and N^- is created as a part of the network, the gain factor of the newly created edge is computed as $g(N^+, N^-)$ = $1 - \text{leak}(x)$, and the edge's capacity is assigned a value of 1. Other nodes in the graph are also assigned the value of 1 for their gain factor as their trust values are the same as their capacities and can be used interchangeably. Initially, the total flow is fixed at unity, so that there is no need for normalization. The computation of trust from the trustor to the trustee has two parts. The first part is to search the path of least distance using breadth-first search and strengthen the path flow. The second part includes two operations, which are, augmenting a flow through the selected path and calculating the residual capacity of each edge from trustee to the trustor, as well as the residual flow that the trustor can send out.

The challenges of this approach are also two-fold. They comprise of the following issues:

i. **Normalization:** Sometimes Graph analogy algorithms result in values that are not in the range that is initially set for trust by the model. Therefore, a proper normalization step needs to be included as part of the model. The reverse mapping is used by RN-trust to find out the value of trust from equivalent resistance, but the model does not specify the range of values of trust and the mechanism of how to guarantee that the final value will lie in that range. Usually, the Network flow-based schemes calculate the maximum flow and perform normalization, to achieve a reasonable result. Advogato does not perform normalization, as it only cares about the rank of each node and not the exact trust value of the node. Thus, designing a valid normalization algorithm is a big challenge graph analogy-based model. Appleseed executes the normalization step for the weight of the edge during the energy diffusion phase. GFTrust avoids normalization by setting the value of initial flow to 1 so as to ensure that the final value of trust lies between 0 and 1.

ii **Scalability:** Generally when dealing with graph analogy-based trust evaluation, the scale of the OSN graph should not be significant. More the scale more will be the space and algorithm's time complexity. It is believed that the combination of the graph simplification and graph analogy based schemes is the best way as first the use of simplification generates a small trusted graph and then graphs analogy cane smoothly applied on the small graph to conduct trust evaluation.

5.7.1.2   COMMON CHALLENGES

We have already discussed the individual challenges faced by the graph analogy and simplification models. Now we will study some of the common roadblocks that the above such approaches based on graphs may face. The most critical four challenges are illustrated below:

5.7.1.2.1   Path Dependence

In a trusted graph, path dependence is caused due to the existence of overlapping trusted paths. It is quite challenging to handle the overlapping nature of trusted paths because they may have several edges in common. A universally accepted solution is yet to be found; models reuse or ignore some of the information provided by the shared edges. TidalTrust selects only the shortest and most reliable paths, and disregard all the other paths. Trust evidence is an essential part of trust evaluation. Thus, an elaborate model is required to handle trust evidence accordingly. It is known from experience that evidence reuse and evidence ignorance both might lead to inaccurate results.

5.7.1.2.2   Trust Decay

Trust is a time sensitive property; therefore, it may decay or change with time. This is known as decay with time. Time can also decay because of iterative recommendations as users have greater trust in their friends than on strangers. This process is known as decay with space. The above two categories of decay imply that time and length of the trusted path should be an indispensable factor of a trust evaluation model. Many factors may incorporate to the process of trust propagation and formulation of user opinion on the OSNs, which include the time of the creation and propagation of information, personality of the user, connection strength, etc.

5.7.1.2.3   Opinion Conflict

Trust is a very subjective property; different users may have differing opinions about the same user or target. One may give a high opinion while some other user might have low opinions about the same node. This phenomenon is known as a conflict of opinion. A big challenge that graph-based approaches face is to how to combine the different opinions. There are several ways to tackle this problem; a ubiquitous technique is to consider taking the paths which have trust levels above a set threshold that is choosing the reliable paths and leaving the others which may cause loss of information. Another way to handle this challenge is to take a WA of all the reliable paths for arriving at a result. To reach to a solution for opinion conflict in trust evaluation we should deeply study two main concepts, one is fully understanding the trustor's features and personal bias; the second is to explore the fundamental principles that form the basis of how people's opinions have formulated and evolved.

5.7.1.2.4   Attack Resistance

Trust evaluation is referred to as a "soft security" technique in comparison to other schemes like encoding. Evaluation systems for trust can be targeted by malicious and selfish users. Some of the attacks in the OSNs are providing lousy service, Sybil attack, bad-mouthing, on-off attack, conflicting behavior attack, and social spamming. There is still a lack of an extensive model that can deal with attacks that occur owing to the open and dynamic nature of OSNs. Thus, a combination of security and trust models are needed to be able to punish users with malicious intent and give incentives to schemes that boost user cooperation.

5.8   ONLINE TRUST EVALUATION IN THE SOCIAL INTERNET OF THINGS (SIOT)

Recently, we have started witnessing the rise of a novel paradigm-the IoT—which describes the idea of everyday physical objects being connected to the internet and being able to identify themselves with other devices. While this fosters the idea of "ambient intelligence," it also introduces new concerns for risks, privacy, and security, owing to which trust evaluation becomes imperative in this environment. As the backbone of IoT is the ability to make decisions without human intervention, trust serves as a crucial factor that can support the establishment of reliable systems and secure connectivity. It could help both the infrastructures and services in managing risks by operating in a regulated manner and avoiding conditions and service failures that could not be predicted.

5.8.1   DEFINITION OF TRUST IN THE SOCIAL INTERNET OF THINGS (SIOT)

Trust in basic terms is defined as the belief that a particular expectation of provision of a service or accomplishment of a task will be fulfilled within a specific context for a specified period. In the SIoT environment, trust is associated with services, applications, devices, systems, and humans. As with most entities, the trust may be evaluated in the context of SIoT in either relative or absolute terms, i.e., in terms of the level of trust or a probability respectively. Similarly, its fulfillment can refer to an action being performed, or the provision of information required.

Trust has many essential characteristics that are used to uniquely determine it in the context of the SIoT. We further observe these characteristics of trust as follows:

- **a. Subjectivity:** The concept of trust is very subjective since it depends much on the perspective of both the trustee and trustor. For example, a certain trustor, say Alice can trust the trustee say Bob while another trustor says Eve may not.
- **b. Asymmetricity:** Trust is not mutual. If Alice trusts Bob for fulfilling a particular trust goal, it does not imply that Bob would trust Alice also to the same degree.

    c. **Dependency on Context:** Trust is highly dependent on the context in the SIoT Space. The context includes the period, environment, and the specific goal to be accomplished. For example, Alice may have trusted Bob for a service previously, but not now.

    d. **Propagative But Not Necessarily Transitive:** Trust may or may not show transitivity, i.e., if Alice trusts Bob and Bob trusts Charlie, it is not necessary that Alice would trust Charlie. However, trust is undoubtedly propagative, i.e., the relationship between Alice and Charlie is reliant on the relationship of trust between Bob and Charlie.

5.8.2 DEPENDABILITY VS. TRUST

When we consider SIoT, we expect that security, reliability, and privacy considerations are not compromised. Any trust to be established in such a digital space must have a certain level of "dependability" in terms of these three factors. Therefore, dependability, representing secure, private, and reliable schemes to deliver quality services or fulfill a specific goal, is a de facto property of a trust-worthy SIoT related digital system. These schemes can further be attributed to variables like availability, safety, confidentiality, and integrity. A thing that is to be kept in mind, though, is that trust and dependability (i.e., a combination of the above mentioned three factors) are not the same things. Trust is an extension of the concept of dependability. Dependability refers to only infrastructure-based confidence as it focuses on the physical systems while trust is an umbrella term, involving within itself a parameter of social capital as well. Social capital refers to the congregation of factors that represent the different aspects of human behaviors, norms, and patterns that have shaped interactions over time. This combination of both social capital and dependability can rightly be used to represent trust as a concept in the digital space, tying both the human and physical nature of things (see Figure 5.5.)

5.8.3 THREE ASPECTS OF TRUST IN SIOT

As mentioned in the previous section, trust is a combination of social relations amongst humans and dependability of physical entities. When applying the same context to the SIoT environment, we can see that trust consists of three major aspects, showing the correlation of the trustor with society (see Figure 5.6.). These three aspects are:

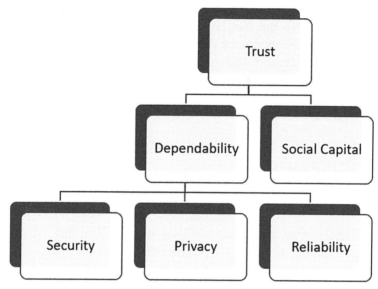

FIGURE 5.5   Concept of trust about dependability and social capital.

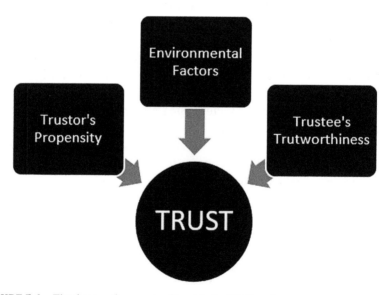

FIGURE 5.6   The three main aspects of trust in the SIoT environment.

a. **Trustor's Propensity:** The term 'propensity' refers to the tendency to behave in a specific manner. In the environment of SIoT, it has great importance due to the generally subjective nature of the concept of trust. The level of trust is considered to have a significant dependency on the behavior and inclinations of the trustor.

b. **Environmental Factors:** Various environmental factors may directly or indirectly affect the level of trust in such a kind of online trust evaluation. These include the vulnerabilities, threats, and risks of the environment in which the work is being done.

c. **Trustee's Trustworthiness:** In general, the reputation of the trustee concerning being counted upon is also a factor that directly influences the trustworthiness of the trustee.

5.8.4   THE CONCEPTUAL TRUST MODEL IN THE SIOT ENVIRONMENT

It is clear now that trust is not a property of either the trustor or trustee but instead, a subjective and asymmetric relationship between the trustor and trustee, characterized by the three main aspects, as mentioned above. Based on these, we may define trust in the SIoT environment in a more comprehensive manner i.e.:

"Trust is the perception of a trustor on a trustee's trustworthiness under a particular environment (within a period), so-called perceived trustworthiness."

A model may thus, be proposed to conceptually demonstrate the concept of trust in the SIoT environment and is illustrated in Figure 5.7.

The model proposed above aggregates both the interactions that take place between the trustor and the trustee and the observations a trustor makes towards a trustee. During these observations and interactions, the risks reflect the different environmental conditions that prevail. Similarly, the trustor' propensity comprises of the requirements of the trust goal as well as his preferences about trustee's trustworthiness. All these factors combine to determine the set of trustworthiness attributes (TAs) that are finally used to derive the perceived trustworthiness and hence, an overall trust value for making a decision. Different forms of weights given to the TAs indicate the trustor's preferences and justify the subjective nature of trust mentioned above, as they may change from trustor to trustor. Therefore, a two-fold goal is achieved when the conceptual model is applied:

   i.   To evaluate the TAs of the trustee's trustworthiness taking into consideration trustor's propensity and environmental factors.

   ii.   To combine these TAs obtained to finally come to a perceived trustworthiness value called the overall trust value.

This also shows that the term 'trust' can be interchangeably be used with the term "perceived trustworthiness."

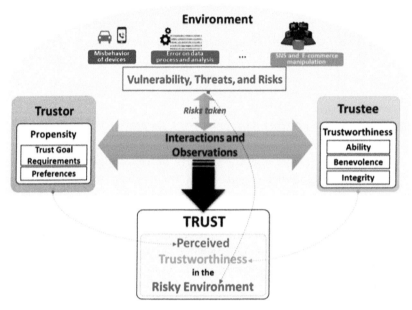

FIGURE 5.7   Conceptual trust model in the SIoT environment. (Reprinted from Truong et al., 2017. http://creativecommons.org/licenses/by/4.0/)

5.8.5 *ONLINE TRUST EVALUATION MODELS*

Trust can never be measured entirely. This is because many factors related to the participants and the environment is complicated to measure or challenging to obtain. To come to the closest possible value of trust that can be evaluated in the context of SIoT, two approaches mainly prevail:

   a.   The direct approach; and
   b.   The prospective approach.

We describe in brief the basis for these approaches in formulating metrics for online trust evaluation:

5.8.5.1 DIRECT APPROACH

The conceptual Trust Model explained in the previous section is also called the direct trust model. It is called so because it uses direct observations on the environment, the trustor and the trustee to evaluate a final value for trust. The procedure can be listed as follows:

    i.   Calculate all TAs of a trustee's trustworthiness.
   ii.   Enlist and take into consideration the preferences and requirements of the individual tasks.
  iii.   Take into consideration the conditions of the environment, including specific threats, risks, and vulnerabilities.
  iv.   Use the factors mentioned above to come to a final value of the trust.

Despite comprehensively including the most influential factors in our calculation of trust via this model, there are still certain inefficiencies in this approach. This is because certain TAs cannot be quantified as they represent qualitative factors like benevolence, integrity, and ability. Measuring them is difficult since it involves ambiguity and variability. Therefore, as we cannot use these TAs, our value of the trust is not entirely indicative of the actual level of trust. Similarly, another disadvantage lies in the complexity of incorporating propensity of the trustor and environmental conditions. Also, reluctance to share personal evidence for trust evaluation is another inherent inefficiency associated with this model. All such key setbacks led to the formulation of another approach which was a more prospective approach.

5.8.5.2 PROSPECTIVE APPROACH

This approach is aimed at removing the hurdles observed in the direct approach of trust evaluation. This approach involves the determination of a set of indicators called trust indicators (TIs). These indicators, as the word implies are used to represent different aspects of trust in a wholesome manner. They could be a TA, a combination of different TAs derived via the Direct Approach or even some other metric based on social interactions and calculated via various Recommendation and Reputation-based models. These TIs can then be summed up (directly or in a weighted manner) to give us the value of computational trust. This computational trust may still not be exactly equal to the value of 'complete' trust. However, it can be persuasively used on behalf of it as it has involved the evaluation of various metrics, not only bounded by the method of calculation suggested by the direct approach. Hence, it is considered a more comprehensive metric for the calculation of trust. The concept of computational trust, comprising of different trust metrics is illustrated in Figure 5.8.

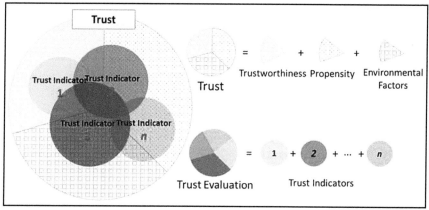

FIGURE 5.8   Computational trust as a combination of multiple metrics. (Reprinted from Truong et al., 2017. http://creativecommons.org/licenses/by/4.0/)

Nevertheless, in the SIoT environment, whichever trust evaluation model we use, it should fulfill two objectives:

i.   Specify a set of indicators or attributes that are indicative of the propensity of the trustor, environmental factors (including risks, vulnerabilities, and threats), and the trustee's trustworthiness.

ii.   Use these indicators to evaluate the value of trust called the computational trust, which will be the closest representative of complete trust and can be used in place of that.

5.9   CONCLUSION

Trust is the most important factor is human social activities, making friends, or making deals; trust plays an important role, thereby increasing the importance of its accurate evaluation. This is also due to the essentiality of maintaining a balance between legitimate public interest and preserving privacy and anonymity. This chapter aimed to provide an overall view of the nature of trusts and its concepts via the various dimensions of online trust evaluation, abreast with the most recent forms of technologies in the digital space. The various approaches to enhancing trust between users may vary with the different environmental factors. In the most recent times, reputation, and recommendation-based feedback frameworks have been developed to comprehensively compute the degree of trust for sellers on e-commerce platforms. Similarly, with the advent of online social networking platforms to meet people and keep in touch, much research is in progress to predict the trustworthiness of people involved. We have explored one of the many approaches to the same, namely Graph-Based Algorithm approaches. Further, to bridge the gap between hardware and software, the IoT concept has also gained a social angle that demands the evaluation of trust. In this manner, we see that indeed, there are open challenges yet to be solved in the domain of this relatively new concept of online trust evaluation. These include privacy preservation, feedback, and testbeds, and tackling distrust. Currently, with the vast amounts of data, trust opinions are hard to formulate given the issue of privacy preservation, making it even more important to balance trust evidence collection and conditions of anonymity. Similarly, current research does not test the performance of existing trust evaluation models, indicating a need for a standard benchmark that takes into consideration representative and standard conditions of the online environment. Lastly, the concept of negative feedback, i.e., distrust also needs to be taken care of, since that may not be propagative and may require different but related methods of trust evaluation. Therefore, with the rapidly changing landscape of Internet Technology, challenges remain evolving, demanding researchers to adapt and continuously come up with newer models and approaches for evaluating trust.

KEYWORDS

- **business to business**
- **directed acyclic graph**
- **internet of things**
- **online social networks**
- **quality of services**
- **trustworthiness attributes**

REFERENCES

Araujo, I., & Araujo, I., (2003). Developing trust in internet commerce. *Proceedings of the IBM Conference of the Centre for Advanced Studies on Collaborative Research* (pp. 1–15). Toronto, Canada.

Backstrom, L., & Jure, L., (2011). Supervised random walks: Predicting and recommending links in social networks. In: *Proceedings of the Fourth ACM International Conference on Web Search and Data Mining* (pp. 635–644). Hong Kong, China, ACM: New York, NY, USA.

Castelfranchi, C., & Falcone, R., (1998). Principles of trust for MAS: Cognitive anatomy, social importance, and quantification. *Proceedings International Conference on Multi Agent Systems (*pp. 72–79*)*. Paris, France.

Dunn, J., (1984). The concept of trust in the politics of John Locke. In: Richard, R. J. B., Schneewind, & Quentin, S., (eds.), *Philosophy in History: Essays on the Historiography of Philosophy (*pp. 279–301*)*. Cambridge University Press.

Golbeck, (2005). *Computing and Applying Trust in Web-Based Social Networks.* PhD Dissertation, University of Maryland.

Gupta, R., Shrivastava, G., Anand, R., & Tomažič, T., (2018). IoT-based privacy control system through android. In: *Handbook of e-Business Security* (pp. 341–363). Auerbach Publications.

Jiang, W. J., Li, F., Wang, G., & Zheng, H., (2015). *Trust Evaluation in Online Social Networks Using Generalized Flow.* IEEE Trans. Comput.

Jiang, W. J., Wang, G., & Zheng, H., (2014). Fluid rating: A time-evolving rating prediction in trust based recommendation systems using fluid dynamics. In: *Proceedings of the 33rd IEEE International Conference on Computer Communications (INFOCOM)* (pp. 1707–1715).

Jiang, W., Wang, G., Md Bhuiyan, Z. A., & Wu, J., (2016). Understanding graph-based trust evaluation in online social networks: Methodologies and challenges. *ACM Computing Surveys, 5,* No. N, Article A.

Jiang, W., Wu, J., & Wang, G., (2015). On selecting recommenders for trust evaluation in online social networks. *Accepted to Appear in ACM Transactions on Internet Technology (TOIT).*

Jiang, W., Wu, J., Wang, G., & Zheng, H., (2015). *Forming Opinions via Trusted Friends: Time-Evolving Rating Prediction Using Fluid Dynamics.* IEEE Trans. Comput. doi: 10.1109/TC.2015.2444842.

Kumari, R., & Sharma, K., (2018). Cross-layer based intrusion detection and prevention for network. In: *Handbook of Research on Network Forensics and Analysis Techniques* (pp. 38–56). IGI Global.

Levien, R & Aiken, A., (1998) Attack-resistant trust metrics for public key certification. *In Proceedings of the 7th USENIX Security Symposium,* 229–242.

Massa, P. A., & Tiella, R., (2005). A trust-enhanced recommender system application: Mole skiing. In: *Proceedings of the 2005 ACM Symposium on Applied Computing (SAC)* (pp. 1589–1593).

Mayadunna, H., & Rupasinghe, P. L., (2018). *A Trust Evaluation Model for Online Social Networks.* National Information Technology Conference (NITC).

Miglani, A., Bhatia, T., Sharma, G., & Shrivastava, G., (2017). An energy efficient and trust aware framework for secure routing in LEACH for wireless sensor networks. *Scalable Computing: Practice and Experience, 18*(3), 207–218.

Oyekunle, R. A., & Arikewuyo, A. O., (2014). *Trust in the E-Market Place: A Case Study of Olx. Ng Users, International Conference on Science, Technology, Education, Arts, Management and Social Sciences iSTEAMS Research Nexus Conference.* Afe Babalola University, Ado-Ekiti, Nigeria.

Radack, S. M., (2011). *Managing Information Security Risk: Organization, Mission, and Information System View.* US Department of Commerce: Gaithersburg, MD, USA.

Sharma, K., & Shrivastava, G., (2014). Public key infrastructure and trust of web-based knowledge discovery. *Int. J. Eng. Sci. Manage., 4*(1), 56–60.

Taherian, A. M., & Jalili, R., (2008). Trust inference in web-based social networks using resistive networks. In: *Proceedings of the 3rd International Conference on Internet and Web Applications and Services (ICIW)* (pp. 233–238).

Truong, N. B., Lee, H., Askwith, B., & Lee, G. M., (2017). Toward a trust evaluation mechanism in the social internet of things. *Sensors, 17,* 1346.

Truong, N. B., Won, T. U., & Lee, G. M., (2016). A reputation and knowledge-based trust service platform for trustworthy social internet of things. In: *Proceedings of the 19th International Conference on Innovations in Clouds, Internet and Networks (ICIN).* Paris, France.

Wang, & Wu, J., (2011). Flow trust: Trust Inference with network flows. *Frontiers of Computer Science in China, 5*(2), 181–194.

Wang, J., Qiao, K., & Zhang, Z., (2018). Trust evaluation based on evidence theory in online social networks. *International Journal of Distributed Sensor Networks, 14*(10).

Wang, Y. D., & Emurian, H. H., (2005). An overview of online trust: Concepts, elements, and implications. *Computers in Human Behavior, 21,* 105–125.

Wang, Y., & Lin, K. J., (2008). Reputation-oriented trustworthy computing in e-commerce environments. *IEEE Internet Computing, 12,* pp. 55–59.

Xianglong, C., & Xiaoyong, L., (2018). In: *Proceedings of the 2018 International Conference on Algorithms, Computing and Artificial Intelligence.* Article No. 23.

Ziegler, &. Lausen, G., (2005). Propagation models for trust and distrust in social networks. *Information Systems Frontiers, 7*(4–5), 337–358.

CHAPTER 6

Trust-Based Sentimental Analysis and Online Trust Evaluation

S. RAKESHKUMAR[1] and S. MUTHURAMALINGAM[2]

[1]Department of Computer Science and Engineering, GGR College of Engineering, Vellore, Tamil Nadu, India, E-mail: rakesherme@gmail.com

[2]Department of Information Technology, Thiagarajar College of Engineering, Madurai, Tamil Nadu, India, E-mail: smrit@tce.edu

ABSTRACT

Sentimental Analysis is a promising frontier in social network big data (BD) mining. Nowadays, e-commerce allows users to buy and sell goods online and enable online feedback. The product sale happens when the customer is satisfied with the reviews of the certified buyers. Therefore, sentimental analysis plays a key role in any prediction system, especially in opinion mining (OM) systems. The instant summarization of product reviews using sentiment analysis (SA) has a great impact in helping the company understand what the customer liked and disliked about the product as well as helps new buyers make an online purchase decision. So the basic task in sentimental analysis is to classify the sentiments as positive, negative, and neutral. It is of importance to decide the trustworthiness of the user who posts the review and also to analyze the review's truthfulness. The degree of impact of the sentiments and the extent to which the reviews are accounted for further selection of the product are identified by using machine learning (ML) techniques such as Naive Bayes (NB), logistic regression (LR), and decision trees (DTs).

There is a need for recommender systems (RS) to provide personalized suggestions of information or products relevant to the user's needs. Though various RSs take into account of the made substantial progresses in theory and algorithm development and have achieved many commercial successes, utilization of the widely available information in online social networks (OSNs) has been largely overlooked. This chapter proposes a framework of implicit social trust and sentiment (ISTS) based RSs, which improves the existing recommendation approaches by applying trust. Thus, a sentiment modeling method is proposed by presenting its modeling rules. The sentiment delivering estimate scheme from two aspects: explicit and implicit sentiment delivering estimate schemes, based on trust chain and sentiment modeling method is implemented. Trust-based calculation of sentiments is done to arrive at the sentimental exploration for future suggestions to the customers to choose a particular product from online e-commerce sites.

6.1 INTRODUCTION

The sentimental analysis also coined as opinion mining (OM) (Bhadane, Hardi, and Heenal, 2015) involves the process of tracking the emotions of the general public about a particular domain or product to arrive at fruitful business or personal decision making. In today's digital world, data is increasing in a drastic manner. So a forum is needed for expressing opinion in various fields like medicine, engineering, marketing, finance, and science. Such forums also act as data source for collecting reviews (Buyya, James, and Andrzej, 2011) and ideas of the public regarding a particular domain.

6.1.1 DATA SOURCES

Anything's success depends on the idea of the people towards the particular thing. Hence, user suggestion plays a vital role in deciding the success story of a particular product or thing. Blogs, Social media, Review sites, and micro-blogging are the popular sources of data for performing sentimental analysis.

6.1.1.1 BLOGS

Blogging has become an important source of information for sentimental analysis. People create blogs in which they express their views about their personal lives and also share posts on various products and issues of latest concern. Thus, blogging helps in identifying the mood of the blogger towards a particular issue thereby providing valuable feedback.

6.1.1.2   SOCIAL MEDIA

Social media (Collomb et al., 2015) sites include Facebook, Twitter, Instagram, LinkedIn, and many more sites which reveal the opinion of the person in an effective manner. These play a vital role in measuring sentiments as they are from public in a leisurely and thoughtful mode. The possibility of linking the various social media technologies act as promising criterion in determining the overall mood of a person towards a particular idea.

6.1.1.3   REVIEW SITES

People always have a curiosity in knowing the idea of another person about a particular product before intending o purchase it, In such a case, in this modern world, they look at the review sites (Devika, Sunitha, and Amal, 2016) for the particular product to decide which item to purchase. For example, to get product reviews sites like www.amazon.com, www.flipkart.cometc is preferred. For various criteria, different sites are referred to obtain maximum benefit.

6.1.1.4   MICRO-BLOGGING

Micro-blogging combines the flavor of instant messaging and blogging which helps in sharing short messages and post that bluntly conveys the polarity of the person who posts the comments.

6.1.2   LEVELS OF SENTIMENT ANALYSIS (SA)

There are three primary levels in the sentimental analysis (Gupta and Vijay, 2009) as shown in Figure 6.1. They are a) Aspect-level SA b) Document-level SA and c) Sentence-level SA.

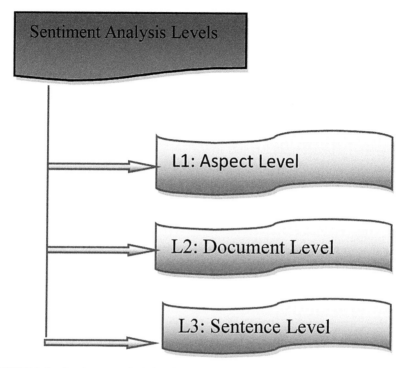

FIGURE 6.1 Sentiment analysis levels.

In Aspect level SA, the classification of the sentiments is based on the specific attributes or aspects. This level provides the minute micro details for performing the sentimental analysis. These terms of positive, negative, and neutral strand. In sentence level, each sentence is subjected to classification of the emotions. The set of sentences constitutes a short document and hence both play a major role in determining the emotions or sentiments at a shallow level. Whereas, the aspect level of SA plays a deeper role in analyzing the sentiments of the entities or objects involved in the scenario.

6.1.3 APPLICATIONS OF SENTIMENT ANALYSIS (SA)

The basic aim of SA is to observe and identify the user opinion on a particular event. The various applications of SA (Fong et al., 2013; Korkontzelos et al., 2016) include online advertising, emotion detection, resource building, hotspot detection, transfer learning, etc.

6.1.3.1   ONLINE ADVERTISING

Using the internet today, online advertising has yielded much income. SA plays a vital role in advertising the various sites with products and also helps in advertising the personal pages of a blogger. This advertising (Gao et al., 2018) brings more customers thereby yielding better profit.

6.1.3.2   EMOTION DETECTION

The sentiment describes the opinion towards a particular thing whereas emotion detection involves the attitude of a person. The SA involves finding the polarity namely positive or negative opinion. The emotion detection (Liu et al., 2015) is concerned with the identification of the various emotions like joy, sorrow, anger, fear, surprise, trust, disgust, and anxiety.

6.1.3.3   RESOURCE BUILDING

Resource building involves building of lexicons and dictionaries which display the emotions based on the positive or negative feedback. The challenges encountered while building the resources are ambiguity in words, granularity (Patel, 2015) and the emotion expression variation among the individuals.

6.1.3.4   HOTSPOT DETECTION

Data availability is huge in internet. So there is a need for some mode through which information could be made available to those who search them. In order to accomplish this, some hotspot forums are organized which helps in easy retrieval of information.

6.1.3.5   TRANSFER LEARNING

This is a cross domain technique for learning which helps in managing the differences of the various domains. This helps in addressing the various techniques like text classification (Pryce, 2010), parts of speech tagging, etc.

6.1.4   TOOLS FOR MEASURING SENTIMENTS

There are various tools for calculating the sentiments of the people. These tools (Ahmed and Ajit, 2015; Bhavish, Sharon, and Nirmala, 2018; John-natan et al., 2017) ease the analysis process by identifying the important features. Some of the tools and their corresponding purpose are tabulated in Table 6.1.

TABLE 6.1   Popular Sentiment Analysis Tools

Sentiment Analysis Tool	Purpose
Bitext	To perform Text Analytics and Concept Extraction
Clarabridge	To perform social media analysis
Hootsuite	To measure social networks
pagelever	To measure Facebook activity
Rapid Miner	To analyze Social Media and perform NLP activities.
VisualText	To Perform NLP and text analytics
Sysomos	Media Analysis Platform which performs social media monitoring.

6.1.5   CHALLENGES IN SENTIMENT ANALYSIS (SA)

There are various challenges in SA (Tayal and Sumit, 2016; Vinodhini and Chandrasekaran, 2012) namely sarcasm which indicates that the author either mentions a positive as positive comment or just tells the negation of it. Parsing acts as an important issue in which there is no clear distinction to which the verb refers to, either subject or object. Poor grammar and spelling also makes it difficult for segregation and classification. Another issue is that what the noun or pronoun actually refers to if there is a dilemma in the reference of a particular context. For example, in the text "I ate and then studied science. It was wonderful," the wonderful means food or studies is the question. So there is a requirement of efficient and effective sentimental analysis techniques for processing the text data and arriving at meaningful detections.

6.1.6   BASIC SENTIMENT ANALYSIS (SA) STRUCTURE

The SA process (Velte, Toby, and Robert, 2010) is depicted in Figure 6.2, which involves fetching input text from either reviews, tweets or any other social media. The text is then subjected to sentiment identification. The feature selection technique (Wen, Wenhua, and Junzhe, 2012) is applied to identify the significant features. The sentiment is finally obtained for the text and the polarity is fixed as positive, negative, or neutral.

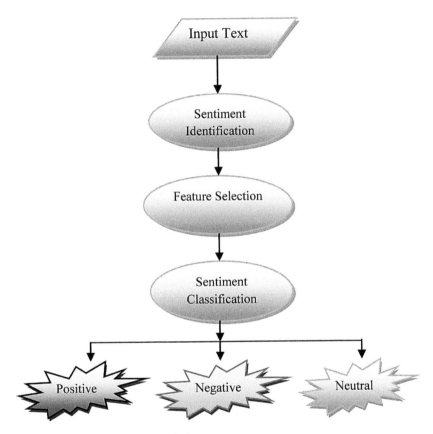

FIGURE 6.2   Basic sentiment analysis structure.

6.2   RELATED WORK

The sentimental analysis plays an important role in product review in recent days. The product review Domain-Aware approach is proposed in (Liu et al., 2015) in which helps in identifying emerging ones, vanishing one, and influential ones by using user trust network. A hypergraph (Vinodhini and Chandrasekaran, 2012) is used to find the time-varying nature of the relationships which overlooks the domain nature of the feature and concentrates on the time-variant for deciding the best product.

Privacy is very important in any e-commerce site. Two major components are user's anonymity and the privacy protection of the website. Both have to be maintained for proper functioning of the system. E-poll system (Bella, Rosario, and Salvatore, 2011) is designed and maintained to ensure privacy. Yet the existing system does not have fairness between the participants of the e-commerce site. So an addition to the existing poll system (Wang and Christopher, 2012) is implemented which targets to maintain privacy in a higher flavor when compared to the existing mechanism.

User feedback plays a major role in many of the online shopping websites like Flipkart, e-bay, and Amazon. The proposed system in (Agjal, Ankita, and Subhash, 2015) uses different dimension of the feedback provided by the user for determining the effective SA based on trust. WordNet (West et al., 2014) is used for the synonym finding of the particular feedback.

Social networks maintain the user profiles for finding the user in the entire group and also find the behavior of the user. So in this regard, the group affinity (Kim and Sang, 2013) is correlated with the user profile to find the relationship bond that exists in the communication. This helps in providing better clarity for the providers based on the user profiles.

Emotions or sentiments help people achieve short or long term objectives. So any arousal or change in the situations tempts the user to react accordingly and thus emotions are predicted. The trust is calculated and if the trust is broken then conflicting emotions (Schniter, Roman, and Timothy, 2015) arise. These emotions force the user to move for recalibrational perspective. This helps in finalizing the sentiments of the user.

Polarity detection is of prime importance in SA. Polarity determines whether the idea of the user is positive, negative, or neutral with respect to the particular criterion. Polarity detection is performed by various SA tools and the results are evaluated for better performance for various applications. Polarity shift (Zirpe and Bela, 2017) is given vital importance as it determines the various security aspects of the sentimental analysis.

6.3 SENTIMENT CLASSIFICATION TECHNIQUES

The sentiment classification is based on three criterions. One is on the basis of the technique deployed for classification. It is subdivided into lexicon (Sonawane and Pallavi, 2017; Wong et al., 2008), statistical, and machine learning (ML) approach. The lexicon is classified into corpus-based and dictionary-based approaches.

The lexicon approach uses the words in the sentence and the lexicons to determine the polarity for particular review content. This is based on semantic and syntax which in-turn depicts the subjectivity. The statistical model assumes mixture of latent features and also includes those from ratings. The next classification is based on the text format (Grobauer, Tobias, and Elmar, 2011) such as aspect basis, document basis and sentence basis. Another category is based on the rating offered by the reviews either as stars or through feature selected for calculating the review. The detailed overall classification of sentiments is depicted in Figure 6.3.

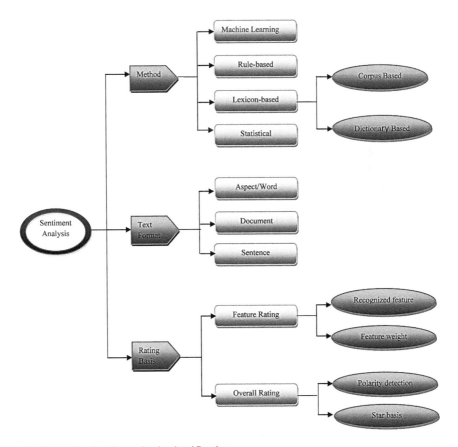

FIGURE 6.3   Sentiment basic classification.

6.3.1 MACHINE LEARNING (ML) TECHNIQUES FOR SENTIMENT ANALYSIS (SA)

Of the various techniques for SA, ML technique (Bhavitha, Anisha, and Niranjan, 2017; Rathi et al., 2018) proves to be a promising one which addresses the various issues of classification of sentiments. The ML technique is broadly classified into supervised and unsupervised learning as shown in Figure 6.4.

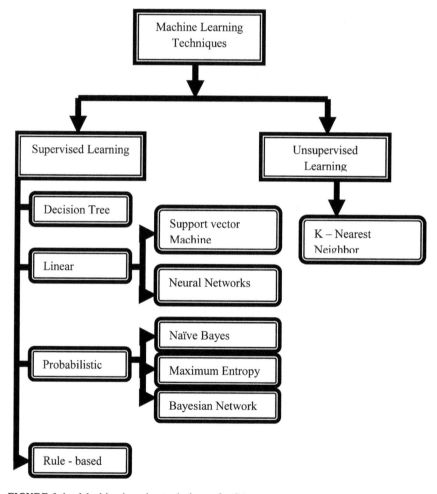

FIGURE 6.4   Machine learning techniques for SA.

6.3.1.1   UNSUPERVISED LEARNING

In unsupervised learning, K-nearest neighbor (KNN) algorithm (Resnick and Richard, 2002) is used for classification of the text. Each entity is classified by using training dataset which maps nearest distance against each object. The major pro is the simplicity in the process of text classification. For a large dataset, segregating the entity type is time-consuming process.

6.3.1.2   SUPERVISED LEARNING

The supervised learning techniques (Rimal, Eunmi, and Ian, 2009) include decision trees (DT), linear, probabilistic, and rule-based classifier. The various algorithms under supervised learning are explained below.

6.3.1.2.1   Decision Tree (DT)

In DT classifier (Shi et al., 1992), the leaf nodes are named with label of the various categories. The internal node is expressed with the attributes selected. The edges from the nodes are named as trial based on weight in the dataset.

6.3.1.2.2   Linear Classifier

For the purpose of classification of the given input data into various classes, linear classifiers are applied. Some of the linear classifiers include neural network (NN) (Kalaiselvi et al., 2017) and support vector machine (SVM). SVM creates hyperplane with maximum distance for the trained sample set. Hence, in addition to speech recognition, SVM (Singh et al., 2018) is applied in text classification also. NN (Mahajan and Dev, 2018) operates with multiple neurons at various levels. The neurons carry the output at each layer. The output of each layer is fed as input to the next layer. The data for the neurons are complex as it is very tough for the correction of values as it has to be back-tracked to various layers behind for correction.

6.3.1.2.3   *Probabilistic Classifier*

This classifier takes each and every class for processing and the probability values are taken for each word. The various probabilistic classifiers include Naïve Bayes, Maximum entropy, and Bayesian Network. NB is simple and it works well with text data. The major problem arises when attributes are selected independently without any assumption. The maximum entropy works along with other classifiers like an exponential classifier in which processing is done by taking input from the dataset and the result is its exponent value.

6.3.1.2.4   *Rule-Based Classifier*

The classifier provides a set of rules with the dataset at the initial stage (Indhuja and Reghu, 2003). In this, the left values are considered as condition and right is the corresponding class labels.

6.4   PRELIMINARIES

The sentimental analysis technique proposed in this work is based on trust and the evaluation of the model through various parameters.

6.4.1   TRUST BASED MODELS

The basic trust model (Rakesh and Gayathri, 2016) is described in Figure 6.5 which involves taking the review comments as input and performs the preprocessing on the input data. The preprocessing includes removal of stop words, POS tagging, and application of parsing. Once the preprocessing is over, it is subjected to SA tool SentiWordNet (Ahmed and Ajit, 2015) and then trust values (Alahmadi and Xiao, 2015; Cao and Wenfeng, 2017) are computed. There are two types' namely implicit trust (Alahmadi and Xaio-Jun, 2015; Cao and Wenfeng, 2017) and explicit trust.

The calculation of polarity (Rahman et al., 2018) of the information obtained happens by using trust factor (Cruz et al., 2018; Panchal et al., 2017). The trust calculation determines the level of each entity giving rating (Anshuman and Misha, 2017) accordingly. There is direct and indirect path in this kind of dealing.

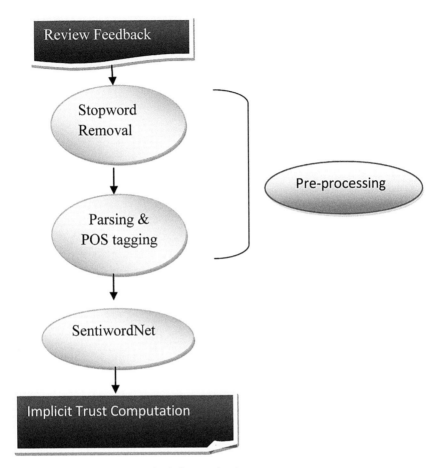

FIGURE 6.5 General trust calculation mechanism.

Direct path (Griffith and Haley, 2018) establish direct relationship between the user A and user B which can be calculated as follows:

$$Tu, v = \frac{P * pos(u,v) + n * pp(u,v) + N * 0}{P + n + N} \qquad (1)$$

where P denotes positive sentiments (El-Qurna, Hamdi, and Mohamed, 2017) from user u to user v (P>0).n denotes the negative sentiments from user u to user v(n<0) and pp denotes the average positive statements from user I to user j. N denotes the neutral sentiments from user u to user v whose value is usually 0.

Figure 6.6 provides the word cloud structure of the mobile review obtained for the year 2018 for the mobiles (Rekha and Williamjeet, 2017) around 15k. The more the frequency of the word, better is the clarity of the word in the structure.

FIGURE 6.6   Mobile review output-SA.

6.4.2   *PROPOSED IMPLICIT TRUST-BASED SENTIMENT ANALYSIS (I-TBSA) ALGORITHM FOR ONLINE EVALUATION*

The implicit trust-based mechanism is devised for easy review management and also for declaration of fruitful, better results. The case study of selecting the best smartphone around 15k is taken as the problem. The reviews and feedbacks are used as a major source for evaluating the problem to arrive at meaningful information. A review of a person about a particular product in multiple social media sites makes his idea a remarkable one for another person to consider. Trust Relation has to be made between the one who gave a review and the one who seek comments.

So trust relation is calculated as follows:

$$Rev = \frac{Rev_{RP,SP}}{Rev_{RP,ASP}} \qquad (2)$$

whereas RP is the review person and SP is the seeking person and ASP is the All Seeking person who wants to know about the best mobile in the market.

$$R = \frac{R_{entry}}{R_{entry} \pm R_{exit}} \qquad (3)$$

Here the R represents the reviews list and R_{entry} represents those who seek review and R_{exit} represents the reviewer who shared idea on the product. Now the trust is defined by the two parameters Rev and R as follows:

$$I\_trust_{RP,SP} = x * Rev + (x-1) * R \qquad (4)$$

I-TBSA Algorithm

Input: Reviews, trust-based values Rev, R
Output: Rating, mean absolute error (MAE)

1. For each review comment RC_i of the user U_i in the list L.
2. If $n(RC_{i(Ui)} > TH)$.
3. Select the reviews RCs & store in a list L1.
4. For U_i in the list L.
5. Tokenize each word in review RC_i.
6. Perform POS Tagging.
7. Calculate polarity $p(RC_i)$.
8. Get probability for each user U_i.
9. Analyze U_i Attitude.
10. End For.
11. End For.
12. For every user U_i.
13. Apply the defined trust parameters for ITrust_fin.
14. ITrust_fin = x * R + (1-x) * Rev.
15. End for.
16. Train ITrust_fin with SVM.
17. Rating prediction from SVM_ITrust_fin.
18. $O = MAE_{ui}$ for each user.
19. Return O/N as the average value.

The algorithm I-TBSA proposes a trust-based technique which helps in performing SA to analyze the attitude of the user towards the issue. By using the trust parameters Rev and R, the new trust namely, ITrust_Fin is calculated and then used for training with SVM model.

6.4.3 CASE STUDY-MOBILE REVIEWS

The smartphones under 15k are considered for the SA to determine the best mobile in that range. Of the various mobiles, the top four mobiles namely Vivo v9, Xiaomi Redmi Note 6, OPPO A7 and Motorola one power are selected for comparison. The reviews of these mobiles are used for identifying the best mobile with respect to the parameters namely memory storage, Camera, and battery (Johnnatan et al., 2017). The features of the four mobiles are compared based on the rating (Poria et al., 2014) (5-star maximum) by the users in the form of reviews in Table 6.2:

TABLE 6.2   Comparison of the User Ratings for the Mobiles Using I-TBSA Approach

Mobile/Feature	Vivo V9	Xiaomi Redmi Note 6	OPPO A7	Motorola One Power
Memory	*****	*****	*****	*****
Battery	***	***	***	****
Camera	****	***	***	****
Processor	***	***	****	****
Polarity based on I-TBSA Algorithm	Positive (90%)	Positive (82%)	Positive (86%)	Positive (95%)

6.5 RESULTS AND DISCUSSION

To evaluate the observed readings for the various mobiles, the parameters such as Precision, recall, and F-measure are calculated. The precision involves fraction of matching reviews to the overall predicted reviews.

$$P = \frac{Matching\ Reviews}{Predicted\ Reviews} \qquad (5)$$

The recall measures a ratio of matching reviews to the actual value.

$$R = \frac{Matching\ Reviews}{Actual\ Value} \tag{6}$$

F-measure value is the combination of precision and recall calculated as follows:

$$F = \frac{2 * P * R}{P + R} \tag{7}$$

The parameters for the various mobiles are depicted in the graphs below.

Figure 6.7 specifies the precision value comparison for the mobiles depicting the moto one power with better values. The precision percentage for Moto one power is 9% ahead of all other mobiles.

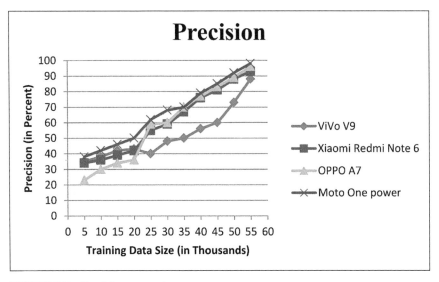

FIGURE 6.7   Precision comparison.

Recall values for all the models are compared in Figure 6.8 and performance of Moto one power is found to be 7% far better than other models.

6.6   CONCLUSION

SA plays a major role in today's world to obtain fruitful results from the existing huge volume of data. The sentimental analysis involves analyzing the polarity of the data as positive, negative, and neutral which helps in analyzing the maximum positive percentage to be the best product in the market. In the proposed work, the various mobiles are compared based on their rating and reviews and the one with maximum rating is given the higher weightage. The proposed system uses a trust-based mechanism with the ML approach for evaluating the best mobile and various parameters like precision and recall are used for evaluation. The Proposed system gives 5% more accuracy than the existing systems.

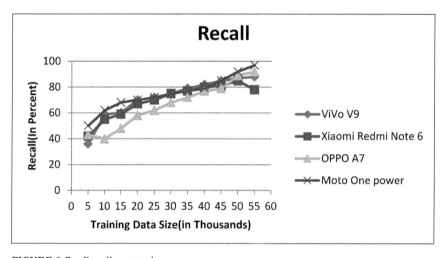

FIGURE 6.8   Recall comparison.

KEYWORDS

- **implicit social trust and sentiment**
- **machine learning**
- **neural network (NN)**
- **online social networks**
- **sentimental analysis**
- **support vector machine (SVM)**

REFERENCES

Agjal, P. M., Ankita, G., & Subhash, T., (2015). "Trust computation using feedback sentiment analysis." *IJSRD-International Journal for Scientific Research and Development, 3*(3), 235–238.

Ahmed, S., & Ajit, D., (2015). "A novel approach for sentimental analysis and opinion mining based on sentiwordnet using web data." In: *International Conference on Trends in Automation, Communication and Computing Technologies, I-TACT.,* 1–5.

Alahmadi, D. H., & Xiao, J. Z., (2015). "Twitter-based recommender system to address cold-start: A genetic algorithm-based trust modeling and probabilistic sentiment analysis." In: *Proceedings-International Conference on Tools with Artificial Intelligence ICTAI International* (pp. 1045–1052).

Alahmadi, D., & Xaio-Jun, Z., (2015). "Improving recommendation using trust and sentiment inference from OSNs." *International Journal of Knowledge Engineering, 1*(1), 9–17.

Anshuman, S. R., & Misha, K., (2017). "A rating approach based on sentiment analysis." In: *Proceedings of the 7th International Conference Confluence 2017 on Cloud Computing, Data Science and Engineering* (pp. 557–562).

Bella, G., Rosario, G., & Salvatore, R., (2011). "Enforcing Privacy in e-commerce by balancing anonymity and trust." *Computers and Security, 30*(8), 705–18.

Bhadane, C., Hardi, D., & Heenal, D., (2015). "Sentiment analysis: Measuring opinions." In: *Procedia Computer Science, 45,* 808–814.

Bhavish, K. N., Sharon, M. J., & Nirmala, M., (2018). "SoftMax based user attitude detection algorithm for sentimental analysis." In: *Procedia Computer Science* (Vol. 125, pp. 313–320).

Bhavitha, B. K., Anisha, P. R., & Niranjan, N. C., (2017). "Comparative study of machine learning techniques in sentimental analysis." In: *Proceedings of the International Conference on Inventive Communication and Computational Technologies, ICICCT* (pp. 216–221).

Bhuiyan, T., Yue, X., & Audun, J., (2008). "Integrating trust with public reputation in location-based social networks for recommendation making." In: *Proceedings-IEEE/WIC/ACM International Conference on Web Intelligence and Intelligent Agent Technology-Workshops, WI-IAT Workshops* (Vol. 3, pp. 107–110).

Buyya, R., James, B., & Andrzej, G., (2011). *Cloud Computing: Principles and Paradigms* (p. 87). Cloud computing: Principles and paradigms.

Cao, J., & Wenfeng, L., (2017). "Sentimental feature based collaborative filtering recommendation." In: *2017 IEEE International Conference on Big Data and Smart Computing* (pp. 463–464). BigComp.

Collomb, A., Damien, J., Omar, H., & Lionel, B., (2015). *"A Study and Comparison of Sentiment Analysis Methods for Reputation Evaluation."*

Cruz, G. A. M. D., Elisa, H. M. H., & Valéria, D. F., (2018). "ARSENAL-GSD: A framework for trust estimation in virtual teams based on sentiment analysis." *Information and Software Technology, 95,* 46–61.

Devika, M. D., Sunitha, C., & Amal, G., (2016). "Sentiment analysis: A comparative study on different approaches." In: *Procedia Computer Science* (Vol. 87, pp. 44–49).

El-Qurna, J., Hamdi, Y., & Mohamed, A., (2017). "A new framework for the verification of service trust behaviors." *Knowledge-Based Systems, 121,* 7–22.

Fong, S., Yan, Z., Jinyan, L., & Richard, K., (2013). "Sentiment analysis of online news using MALLET." In: *Proceedings-International Symposium on Computational and Business Intelligence* (pp. 301–304). ISCBI.

Gao, J., Hua, Y., Limin, W., & Yu, Q., (2018). *Evaluating Sentiment Similarity of Songs Based on Social Media Data*, 1–6.

Griffith, A. N., & Haley, E. J., (2018). "Building trust: Reflections of adults working with high-school-age youth in project-based programs." *Children and Youth Services Review*.

Grobauer, B., Tobias, W., & Elmar, S., (2011). "Understanding cloud computing vulnerabilities." *IEEE Security and Privacy, 9*(2), 50–57.

Gupta, R. S. A., & Vijay, A., (2009). "Cloud computing architecture." *Chinacloud. Cn, 40*, 29–53.

Hussein, D., & Mohey, E. D. M., (2018). "A survey on sentiment analysis challenges." *Journal of King Saud University-Engineering Sciences, 30*(4), 330–338.

Indhuja, K., & Reghu, R. P. C., (2003). "Fuzzy logic based sentiment analysis of product review documents." In: *2014 1ˢᵗ International Conference on Computational Systems and Communications, ICCSC 2014* (pp. 18–22).

Johnnatan, M., Joao, P. D., Elias, S., Miller, F., Matheus, A., Lucas, B., Manoel, M., & Fabrıcio, B., (2017). "An evaluation of sentiment analysis for mobile devices." *Social Network Analysis and Mining*.

Jose, M. G. G., Víctor, M. R. P., María, D. L., Juan, E. G., & Effie, L. C. L., (2018). "Multimodal affective computing to enhance the user experience of educational software applications." *Mobile Information Systems*, pp. 1–10.

Kalaiselvi, N., Aravind, K. R., Balaguru, S., & Vijayaragul, V., (2017). "Retail price analytics using back propogation neural network and sentimental analysis." In: *2017 4ᵗʰ International Conference on Signal Processing, Communication and Networking, ICSCN* (pp. 1–6).

Kim, M., & Sang, O. P., (2013). "Group affinity-based social trust model for an intelligent movie recommender system." *Multimedia Tools and Applications, 64*(2), 505–516.

Korkontzelos, I., Azadeh, N., Matthew, S., Abeed, S., Sophia, A., & Graciela, H. G., (2016). "Analysis of the effect of sentiment analysis on extracting adverse drug reactions from tweets and forum posts." *Journal of Biomedical Informatics, 62*, 148–158.

Li, X., Xiaodi, H., Xiaotie, D., & Shanfeng, Z., (2014). "Enhancing quantitative intra-day stock return prediction by integrating both market news and stock prices information." *Neurocomputing., 142*, 228–238.

Liu, B., & Lei, Z., (2012). "A survey of opinion mining and sentiment analysis." In: *Mining Text Data* (pp. 415–463).

Liu, S., Cuiqing, J., Zhangxi, L., Yong, D., Rui, D., & Zhicai, X., (2015). "Identifying effective influencers based on trust for electronic word-of-mouth marketing: A domain-aware approach." *Information Sciences, 306*, 34–52.

Mahajan, D., & Dev, K. C., (2018). "Sentiment analysis using RNN and Google translator." In: *Proceedings of the 8ᵗʰ International Conference Confluence on Cloud Computing, Data Science and Engineering, Confluence* (pp. 798–802).

Medhat, W., Ahmed, H., & Hoda, K., (2014). "Sentiment analysis algorithms and applications: A survey." *Ain. Shams Engineering Journal, 5*(4), 1093–113.

Panchal, V., Zaineb, P., Sneha, P., Rhea, S., & Reena, M., (2017). "Sentiment analysis of product reviews and trustworthiness evaluation using TRS." *International Research Journal of Engineering and Technology*.

Patel, D., (2015). *"Approaches for Sentiment Analysis on Twitter: A State-of-Art Study."*

Poria, S., Erik, C., Grégoire, W., & Guang, B. H., (2014). "Sentic patterns: Dependency-based rules for concept-level sentiment analysis." *Knowledge-Based Systems, 69*, 45–63.

Pryce, R., (2010). "Choosing the right manikin for the job." *Occupational Health and Safety (Waco, Tex.), 79*(6), 60–62.

Rahman, F., Hamadah, T., Wan, A., Shah, N. S. H., Wida, S. S., & Gyu, M. L., (2018). "Find my trustworthy fogs: A fuzzy-based trust evaluation framework." *Future Generation Computer Systems.*

Rakesh, S. K., & Gayathri, N., (2016). "Trust based data transmission mechanism in MANET using sOLSR." *Annual Convention of the Computer Society of India*, 169–180.

Rathi, M., Aditya, M., Daksh, V., Rachita, S., & Sarthak, M., (2018). *Sentiment Analysis of Tweets Using Machine Learning Approach*, 1–3.

Rekha, and Williamjeet, S., (2017). "Sentiment analysis of online mobile reviews." In: *Proceedings of the International Conference on Inventive Communication and Computational Technologies, ICICCT* (pp. 20–25).

Resnick, P., & Richard, Z., (2002). "Trust among strangers in internet transactions: Empirical analysis of EBay' s reputation system." *Advances in Applied Microeconomics*, 127–157.

Rimal, B. P., Eunmi, C., & Ian, L., (2009). "A taxonomy and survey of cloud computing systems." *NCM 2009–5*[th] *International Joint Conference on INC, IMS, and IDC*, pp. 44–51.

Schniter, E., Roman, M. S., & Timothy, W., (2015). Shields. "Conflicted emotions following trust-based interaction." *Journal of Economic Psychology, 51*, 48–65.

Shi, D., Gongping, H., Shilong, C., Wensheng, P., Hua-Zhong, Z. Z., Dihua, Y., & Mien-Chie, C. H., (1992). "Overexpression of the C-erbB-2/Neu–encoded P185 protein in primary lung cancer." *Molecular Carcinogenesis, 5*(3), 213–218.

Singh, P., Ravinder, S., Sawhney, & Karanjeet, S. K., (2018). *"Sentiment Analysis of Demonetization of 500 & 1000 Rupee Banknotes by Indian Government"*, *4*(3), 124–129.

Sonawane, S. L., & Pallavi, V. K., (2017). *Extracting Sentiments from Reviews: A Lexicon-Based Approach a Sammplate for Information*, 38–43.

Sosinsky, B., (2010). *Cloud Computing Bible* (p. 762). Wiley Publishing Inc.

Taboada, M., Julian, B., Milan, T., Kimberly, V., & Manfred, S., (2011). *"Lexicon-Based Methods for Sentiment Analysis,"* *37*(2), 267–307.

Tayal, D. K., & Sumit, K. Y., (2016). "Fast retrieval approach of sentimental analysis with implementation of bloom filter on Hadoop." In: *2016 International Conference on Computational Techniques in Information and Communication Technologies, ICCTICT-Proceedings* (pp. 14–18).

Velte, A. T., Toby, J. V., & Robert, E., (2010). *Cloud Computing : A Practical Approach.*

Vinodhini, G., & Chandrasekaran, R. M., (2012). "Sentiment analysis and opinion mining: A survey." *International Journal of Advanced Research in Computer Science and Software Engineering, 2*(6), 282–292.

Wang, S., & Christopher, D. M., (2012). *"Baselines and Bigrams: Simple, Good Sentiment and Topic Classification, 2*, 90–94.

Wen, B., Wenhua, D., & Junzhe, Z., (2012). "Sentence sentimental classification based on semantic comprehension." In: *Proceedings-2012 5*[th] *International Symposium on Computational Intelligence and Design* (Vol. 2, pp. 458–461). ISCID.

West, R., Hristo, S. P., Jure, L., & Christopher, P., (2014). *"Exploiting Social Network Structure for Person-to-Person Sentiment Analysis,"* 1409–2450.

Wong, W. K., Sai, O. C., Tak, W. Y., & Hoi, Y. P., (2008). "A framework for trust in construction contracting." *International Journal of Project Management, 26*(8), 821–829.

Zirpe, S., & Bela, J., (2017). "Polarity shift detection approaches in sentiment analysis: A survey." In: *Proceedings of the International Conference on Inventive Systems and Control, ICISC* (pp. 1–5).

CHAPTER 7

Denaturing the Internet Through a Trust Appraisement Paradigm by QoS Accretion

SHERIN ZAFAR and DEEPA MEHTA

Department of CSE, SEST Jamia Hamdard, India,
E-mails: zafarsherin@gmail.com (S. Zafar),
*   deepa.mehta12@gmail.com (D. Mehta)*

ABSTRACT

The criterion of trust is an important aspect of instilling quality of service (QoS) in the Internet world today. Web services (WS) nowadays are independent, self-contained, and work on disparate platforms, performing tasks that can be simple requests as well as can be the complicated type of business processes and applications. Trust is a personalized service reflecting an individual's opinion on the QoS. Better is the QoS much better will be the trust paradigm and vice-versa. The internet we-services in data transfer face various security issues of authentication, non-repudiation, confidentiality, etc. Various attacks like black-hole, jamming, flooding, and denial of service (DoS) holds a question mark on QoS, thus loosing trust of user. The new-fangled trust appraisement paradigm evaluated and discussed in this research chapter focuses on QoS accretion, thus denaturing internet through trust. The new-fangled trust-based approach maintains a conviction score for each hub along with the peer-to-peer networks. The coordinator process checks each hubs trust value and generates a key and distributes it across the peer-to-peer web, with trust value more or equal to the threshold value. The path of data transfer with the highest trust value is then selected for transmission. The proposed approach discussed in this chapter demonstrates that trust should be adjusted and improved by tracing the fundamental issues that are being tended in the present usage work.

7.1   INTRODUCTION

In recent years internet has become a vital piece of our day-to-day lives. With no limitation to access and utilization of cell phones, the internet usage is crossing all its boundaries. As web-based life advances quickly, the standards of utilization ceaselessly change. Trust has a vital task to carry out in questionable conditions and it is available in human cooperation. Trust frequently mirrors the desire that individuals are kind and carry their work rigorously. It can unquestionably be felt that long-range interpersonal communication administrations require trust, so as to work appropriately. There are obviously a few parts of person to person communication benefits that request some trust to be maintained. Individuals regularly share their private data in long-range interpersonal communication benefits that they essentially don't need the entire world to see. Web-based social networking clients need to survey whether to trust or not to confide in the administration. It is fascinating to recognize what influences trust development in internet-based life. For instance, how does an individual's behavior normal for the client, stage qualities, or associations with the different client are questions that impact trust arrangement.

There could be a few factors that lead to expansion of trust or impact adversely otherwise. Trust could likewise be all the more effectively framed in specific circumstances. The upcoming sections 1.2 and 1.3 of this chapter focus the background study of trust followed by research problems and findings. Sections 1.4, 1.5, and 1.6 specify definitions and measurements, different facets and various properties of trust. Section 1.7 details various models of trust followed by trust appraisement paradigm by QoS accretion in Section 1.8. Section 1.9 discusses the conclusion and future prospects of the approach discussed in this chapter.

7.2   BACKGROUND OF STUDY

The thought for this examination emerged from past researches where the precursors of innovation of trust and its apparent results were contemplated. The thought was to concentrate on web-based networking media and to think about its trust precursors as internet utilization has hugely expanded in the recent years. Past examinations about online trust mostly center on online data sources and their correspondence. In past years, a few examinations about trust and security issues with online life have been made. Most past investigations have concentrated on utilizing trust to disclose specific administration issues. A few investigations have concentrated on social capital in long range interpersonal communication administrations. Social capital includes trust and could be framed because of huge utilization and encouragement of internet in current scenario.

Trust has connected with utilization in numerous web administrations. It has high questionable conditions due to desires for fruitful exchanges. It has ended up being an impetus in buyer advertiser connections. The primary explanation for buyers, not to take part in web-based business is believed to be absence of trust. A web retailer who neglects to demonstrate reliability might be seen as deft and exploiting the web foundations (Alesina and La Ferrara, 2002). Authors have contemplated development of "cyber-trust" through clarifications of social determinants that mold trust in the web (Artz and Gil, 2007). Recent works focused on concerns of web protection around (Dwyer et al., 2007) informal communications among different individuals through long ranges of interpersonal communications.

Researchers have concentrated on how trust influences on building new connections on the web, considered the connection between relational trust and authors (Barnes, 1954; Brunie, 2009) and also considered the innovation supplier and the data framework in social area advances. Related work also (Buskens, 1998) considered elements influencing the acknowledgment of long-range informal communication locales with the all-encompassing trust association and management (TAM) display, which incorporates trust and its associated hazards.

There is a great deal of material accessible on the points of trust and internet-based life independently. In the course of recent years, there have been 37,466 logical distributions on the subject of web-based life and the numbers have been boundlessly developing.

Productions with respect to both online networking and trust have quickly expanded in the course of recent years through internet-based life distributions (991 productions in the course of recent years). There has not been an investigation that has endeavored to clarify the procedure of how trust in long-range informal communication administrations is framed and what are the significant givers. The forerunners of trust in long-range interpersonal communication administrations would intrigue to know how trust has been found to achieve security goals. Knowing components foresee trust usage by an intriguing new person to person communication administrations persistently (Wikipedia, 2015). The forthcoming areas of this chapter center and examine about security issues, assaults, trust, trust assessment, and QoS ideas. Figure 7.1 shows social dependability of trust in today's scenario.

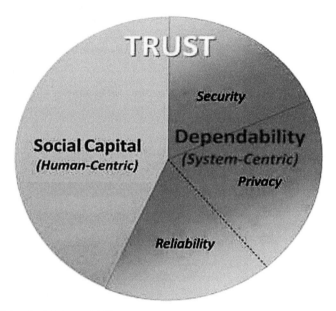

FIGURE 7.1   Social dependability of trust in today's scenario.

7.3   RESEARCH PROBLEM AND QUESTIONS

There may be a few things affecting person to person communication benefits of clients towards the administration of trust. Clients need to gauge security concerns, the stage supplier's inspiration, and ability to ensure security of their data. Social dangers may emerge from contacts whether they are new, online colleagues or closer companions.

Clients can't know everything, so they need to confide dependency on something. The target of this research chapter is to figure out what are the forerunners of trust in long-range interpersonal communication administration. The point is to look for a clarification for trust arrangement in interpersonal interaction administrations. In view of the destinations of exploration, the fundamental research questions are:

1. What are the precursors of trust in interpersonal interaction administrations?
2. How do trust forerunners influence trust towards long range informal communication administrations?

The main inquiry looks for distinct answers that center's around recognizing the predecessors and the purposes for trust. The potential factors behind trust in long-range interpersonal communication administrations are looked into through applicable writing as depicted in Figure 7.2. The second inquiry looks for affirmation and clarification of the significance and impact of every predecessor. The significance of every forerunner is experimentally evaluated. The outcomes could intrigue organizations growing new long-range interpersonal communication administrations. Substantial associations may likewise profit by the outcomes that they want to present inside long range interpersonal communication administrations.

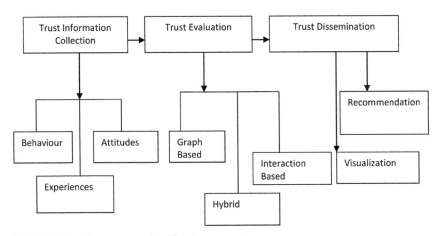

FIGURE 7.2   Trust system classification concepts.

The electronic era gave way to the multi-disciplinary idea of trust. It is consequently not astonishing that there are as of now some survey articles on various parts of trust. This area intends to display an outline of existing survey articles on trust that allude to trust frameworks, notoriety frameworks, and appraisal of trust. Current audit articles are classified from the perspective of trust utilizations and arrangement of its applications, specifically, shared (P2P) applications, and internet applications. Related work on specially appointed system overviews on its trust plans, its objectives, its highlights and structure of trust framework (Azer et al., 2008). Previous studies (Momani and Challa, 2010) audit the treatment of trust explicitly in remote and sensor spaces.

Authors have also (Suryanarayana and Taylor, 2004) audited trust in P2P applications into three general classes: trust certification and arrangement based, trust notoriety based, and trust informal community-based. The essential thought at the level of accreditation and arrangement is utilization of certifications that empower access control of assets. These strategies help along suspicion of certain trust on the asset proprietor. The trust notoriety based framework, interestingly gives a capacity to assess the trust of the asset proprietor dependent on notoriety esteems. The informal community-based strategy utilizes social connections for ranking hubs in the system. Authors have constrained their work (Suryanarayana and Taylors, 2004) to three diverse interpersonal organization based strategies for assessing trust: network-based notoriety (Yu and Singh, 2000a), lament strategy (Sabater and Sierra, 2002), and hub positioning approach (Sabater and Sierra, 2002).

Internet applications have examined trust along three different angles based on contents of website, application of web and its administration. Authors have (Beatty et al., 2011) performed a meta-investigation of customer trust on web-based business locales that concentrate on the association of the website substance for providing reliability.

Recent works (Grandison and Sloman, 2000) outlay trust along with perspectives of distinguishing its requirements for the application of business and to give an extensive arrangement for engineering trust. Work has also done to issue trust along with the semantic web for the advancement of locales and its applications for exploring trust from the viewpoint of software engineering (Artz and Gil, 2007). Trust has also explored the zone of administration; registering where it chooses the clients best administration framework (Malik et al., 2009; Chang et al., 2006). Researchers have methodically surveyed trust along with its notoriety framework for benefit of web service and to propose a topology arrangement along measurements of concentrated/decentralized, specialists/assets and worldwide/ customized (Wang and Vassileva, 2007; Josang et al., 2007).

Trust has also been reviewed as a rule that outlines on the idea of trust board, models, and data models (Ruohomaa and Kutvonen, 2005; Golbeck, 2006). Trust has also being surveyed along the contents of web including webpage counting, sites counting, information of semantic web, web page administration in P2P systems and their applications.

Some writings of researchers audit concentrate on strengthening trust against assaults which is now an imperative part of security framework (Hoffman et al., 2009). Assaults and its barrier instruments are sorted as *self-advancing, *whitewashing, *defaming, *arranged, and *forswearing. Researchers specify that trust frameworks are powerless against assaults of deceiving procedures that occur from various unscrupulous numbers (Kerr and Chon, 2009). Examination of unique trust assaults is done on its notoriety frameworks utilizing standard procedures against hypothetical investigations of trust (Josang and Goldbeck, 2009).

These audit articles centers trust on perspective of software engineering and not on a sociology viewpoint. Trust models are generally presented on grounds of internet applications that are produced and contemplated on software engineering paradigm. New trust investigations are raised on grounds of interpersonal organizations, informal organizations and on society matters.

The rise of interpersonal organizations has prodded new research in the investigation of trust and, as of late, various trust models for informal organizations have been created with an explicit spotlight on social parts of trust. This chapter presents a far reaching survey of trust paradigm analysis in interpersonal organizations covering both software engineering and sociologies as depicted in Figure 7.3.

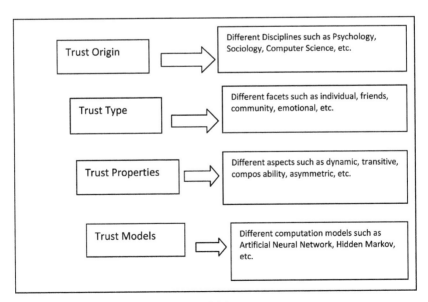

FIGURE 7.3   Trust measurements and definitions.

7.4   DEFINITIONS AND MEASUREMENTS OF TRUST

Human social connections acknowledge trust as a noteworthy segment of security analysis. Trust is specified as a certainty proportion carried out by an element regardless of absence of controlling methodology (Singh and Bawa, 2007). Trust is represented by different definitions in brain and human science and interpreted differently in software engineering. The following definitions of trust specify its analysis in different fields:

1. **Trust in Psychology:** Brain science views trust as a mental condition of the person, where the trustor dangers being helpless against the trustee and desires for the trustee's expectations or conduct (Rotter, 1967; Tyler, 1990; Rousseau et al., 1998). Trust is considered to have three viewpoints: intellectual, emotive, and social.

2. **Trust in Sociology:** Humanism characterizes trust as a wager about the unforeseen activities of the trustee. Individual and society are the two perspectives through which trust is considered. Individual perspective or dimension of brain research considers trustors weakness as the main cause (Dumouchel, 2005; Sztompka, 1999; Jhaver, 2018). Trust is separated from collaboration within the sight of confirmation (an outsider directing the communication and giving authorizations if there should arises an occurrence of trouble).

    Collaboration within the sight of the shadow of things to come (i.e., dread of future activities by the other party) is viewed as trust. In this regard, social trust has just two features, intellectual, and conduct, with the emotive viewpoint working after some time as trust increments between two people. Society level views trust as a social gathering property spoken by the mental conditions of group. Social trust infers that individuals from a social gathering act by the desire that different individuals from the gathering are additionally reliable (Lewis and Weigert, 1985) and anticipate trust from other gathering individuals. So at the level of society trust incorporates institutional framework.

3. **Trust in Computer Science:** Software engineering groups the trust paradigm into two main classes of "client" and "framework" (Marsh, 1994; Mui, 2003). Brain research gave rise to client trust. Researchers define trust as a "substance of abstract desire on another's future conduct, suggesting trust as a natural phenomenon." This suggests trust is naturally customized. Companies like

Amazon and eBay utilizes past collaborations of individuals as a trust dependence parameter. Social trust is the cooperation of one another through dependency experience. Trust is incremented between individuals with surety of experience and diminishes if the relationship experience is false. Online trust framework is viewed through coordinate trust, suggestion trust and proposal trust. Immediate experience of part with other party is specified as coordinate trust. Encounters of different individuals of informal community with the other party relates to suggestion trust. Proposal trust depends on the propagative property of trust. In P2P-based trust models, individuals accumulate data about different individuals utilizing their informal communities, now, and then additionally alluded to as referral systems. Each companion models differs trust as: their reliability as connection accomplices (ability to give administrations) called mastery, and their dependability as recommenders (capacity to give great proposals) alluded to as friendliness.

After every association in nature, the aptitude of the communication accomplice and the friendliness of the companions in the referral affix that the cooperation are refreshed to mirror the experience of the part in that connection. The quick neighbors of the part are additionally intermittently refreshed to mirror the changes in the assessed trust of those individuals through neighbor's aptitude and amiability. The standard idea of "trust," got from the security area (Yao et al., 2010), is the desire that a gadget or framework will loyally carry on in a specific way to satisfy its proposed reason.

For instance, a PC is reliable if its product and equipment can be relied upon to execute, with the end goal that its administrations are as yet accessible today, unaltered, and carry on in the very same route as they did yesterday (Moreland et al., 2010). The idea of framework "trust" is upheld by both programming and equipment based arrangements (Seshadri et al., 2004) displayed by a product based instrument, or (Chen and Li, 2010) depicted through an equipment-based system.

Framework "trust" is the core extent of this study. In any case, it is noted that the trusted computing group (TCG), incorporates a trusted platform module (TPM), a cryptographic microcontroller framework empowers the TPM to be a base of trust for approving both the equipment and programming qualities of a remote PC on which the TPM is introduced. Two essential perspectives describe a trust relationship: hazard and reliance, the wellspring of hazard is the vulnerability with respect to the goal of the other party. Trust relationships established through the connection interest of two gatherings and the relationship cannot be accomplished without the dependence of each other. The relationship isn't a trust relationship if connection and dependence don't exist. Since hazard and association are essential conditions for trust changes, the span of a relationship adjusts both the dimension and the type of that trust (Rousseau et al., 1998).

7.5   DIFFERENT FACETS OF TRUST PARADIGM

This section discusses the different facets of trust from various disciplines:

1. **Calculative:** This trust characterizes consequences figuring from the benefit of the trustor, intended to augment its stakes in the communication. This part of trust is pervasive in financial aspects, where the detainee's predicament diversions are utilized to demonstrate trust and collaboration. It is additionally normal in authoritative science. An objective performer will put trust on the off chance specifying that the proportion of the possibility of gain to the shot of misfortune is more noteworthy than the proportion of the measure of the potential misfortune to the measure of the potential gain (Coleman, 1990).

2. **Relational:** This characterizes trust developed after some time because of rehashed cooperation's between the trustor and trustee. Data accessible to the trustor from the relationship itself shapes the premise of social trust. Trustee's goals are assumed through quality of past cooperation (Rousseau et al., 1998) named as coordinate trust specified in software engineering. It relates trust dependence specified by direct communications between the two gatherings. Researchers have contemplated calculative and social trust association with specification of change of work (Scully and Preuss, 1996). For representatives working in the conventional setting of tight individualized errands, calculative trust assumes a more noteworthy job than social trust by settling on a choice to take up courses supporting work change. Interestingly, for representatives working in group based situations, the choice depends on social trust (of the partners) as opposed to calculative trust (i.e., singular gain).

3. **Emotional:** The enthusiastic part of trust characterizes the security and solace in depending on a trustee. In brain research, enthusiastic trust is seen to be a result of direct relational connections. Passionate trust influences intellectual trust by impacting the trustors positive impressions of their progressive relationship. Exact investigations (Taylor, 2000) demonstrate that the trustor's past direct encounters can influence his/her feelings and henceforth his/her passionate trust on the trustee (Holmes et al., 1991).

4. **Cognitive:** This intellectual part of trust is characterized being dependent on reason and levelheaded conduct. As indicated by the social capital hypothesis, three types of capital can influence intellectual trust: data channels, standards sanctions, and the trustee's commitments to the trustor. Furthermore, the social connection structures inside systems and the quality of the ties between individuals can affect the trustor's intellectual trust on the trustee. Positive referrals inside the relations of informal communities increment subjective trust in the trustee. Subjective trust is followed trust and enthusiastic trust leads to development of great negative desires against the trustee. Passionate trust is compared to enthusiastic security model that empowers to go beyond the accessible proof by feeling guaranteed and agreeable on a trustee (Mollering, 2002; Barka, et al., 2018).

5. **Institutional:** This trust specifies a situation that empowers collaboration among individuals and punishes mischievous activities. Such backings can exist at authoritative dimension and at the societal dimension, for example, legitimate framework secures individual rights and property. Publius (Waldman et al., 2000) is an application that utilizes institutional trust to secretly distribute materials through an extent that controlling and messing production becomes extremely troublesome.

6. **Dispositional:** This trust perceives that through the span of their lives, individuals create summed up assumptions regarding the dependability of other individuals. This is epitomized in dispositional trust, which is reached out to everybody paying little respect to whether the trustor knows about the reliability about the trustee or not. Agreeable inclination is a type of social capital and hereditary inclination that can be considered to be added to dispositional trust.

troublesome situation if data is opposing. Trust composability gives a method for processing trust in informal organizations (Golbeck, 2005). An idea based on transparently characterizing trust creation capacity is used to access trust dependability on trust composability (Richardson et al., 2003). Trust arrangement framework is dependent on the structure of its connections. Trust models spread the trust element to additionally utilize and highlight trust esteems through few trust chains that settle on trust choices.

6. **Subjective Trust:** Trust is subjective as it is abstract. For instance, Bob gives a feeling about a motion picture. In the event that Alice believes Bob's assessments are in every case great, she will confide in Bob's audit. Be that as it may, John may contemplate Bob's sentiments and may not confide in the audit. The emotional idea of trust prompts personalization of trust calculation, where the inclinations of the trustor directly affect the processed trust esteem. Different trust models think about personalization of trust. In software engineering, customized trust contemplates that an individual is probably going to have distinctive dimensions of trust according to other people, with respect to the implanted informal organization.

7. **Asymmetric Trust:** Trust is commonly hilter kilter. A part may confide in another part more than it's trusted back. In any case, when the two gatherings are reliable, they will unite to a high common trust after rehashed associations. On the other hand, in the event that one of the individuals does not act in a reliable way, the other part will be compelled to punish him/her, prompting low shared trust. Asymmetry can be viewed as an uncommon instance of personalization. Asymmetry contrasts in people groups, observations, conclusions, convictions, and desires. The uneven idea of trust has been recognized in different chains of importance inside different associations.

8. **Self-Strengthening Trust:** Trust is self-strengthening; individuals act emphatically with different individuals whom they trust. Correspondingly, if the trust between two individuals is beneath some edge, it is exceptionally impossible that they will connect with one another, prompting even less trust on one another. This part of trust has gotten similarly less consideration in the writing.

9. **Event Touchy Trust:** Trust sets aside a long opportunity to assemble, however, a solitary high-affect occasion may crush it totally. This part of trust has received less consideration in software engineering.

7.7   MODELS OF TRUST

This section focuses on various trust models. The models of trust utilize comprehensive measures of machine learning (ML) procedures, systems based on heuristic paradigms and strategies that are conduct based. Factual ML procedures center's around giving a sound scientific model to confide in the trust model. Heuristics methods center with respect to characterizing a handy model for actualizing vigorous trust frameworks. Conduct models center with respect to client conduct in the network. Bayesian frameworks (Mui et al., 2002; Josang and Ismail, 2002; Vianna and Priya, 2018) and convection models (Josang, 2001; Yu and Singh, 2002b; Josang et al., 2006) are the significant instances of simply factual strategies. Normally in Bayesian frameworks, paired evaluations (genuine or deceptive) are utilized to survey trust by measuring refreshed beta thickness capacities.

Conviction framework demonstrates purchaser's conviction in regards to reality of rating through trust calculation. Each system differ in their conviction; for instance, Dempster-Shafer hypothesis utilizes an abstract rationale (Yu and Singh, 2002a; Josang, 2001; Josang et al., 2006). Different arrangements dependent on ML algorithms like; artificial neural networks (ANNs) and hidden Markov models (HMMs) for calculation of trust.

HMM is used as an assessing mechanism for developing recommender trust models that propose a discrete trust display. Factual and ML arrangements intricately move towards heuristic methodologies for trust incorporation. These arrangements characterize a functional, vigorous, and straightforward trust framework. Many researchers have presented a trust arrangement comprising of key heuristics-based factual HMM model for notoriety appraisal. In conduct based trust models trust is assessed on the conduct of individuals and interpersonal organizations (Song et al., 2004; ElSalamouny et al., 2010; Xiong and Liu, 2004; Huynh et al., 2006; Malik et al., 2009; Adali et al., 2010).

Social trust is assessed through two kinds of trust: discussion trust and engendering trust. Discussion trust determines to what extent and how as often as possible two individuals speak with one another. Longer and increasingly, visit correspondence shows more trust between the two gatherings. Proliferation/Engendering trust alludes to the spread of data as data acquired from one part to different individuals shows that a high level of trust is being set on the data and verifiably its source.

7.8   PROPOSED TRUST APPRAISEMENT PARADIGM BY QoS ACCRETION

One of the significant security issues in web today is the bad conduct of hubs around the infrastructure-less networks. They can publicize themselves of having a most brief course to the goal to transmit the information and might quit sending the information sooner or later leading to loss of parcels. Cell phones speak with one another through a multi-bounce course, utilizing collaborating mediator hubs. An abnormal state of participation is basic for applications that require continuous information transmission; for example, fighters handing-off data in a front line. Be that as it may, the constrained vitality supply of cell phones raises questions about the capacity of each hub to be completely helpful. Therefore, bundle conveyance can't be ensured notwithstanding when vindictive hubs are absent and resending information parcels does not give a decent arrangement.

On the off chance that malevolent hubs are available, they may endeavor to lessen arrange availability (and in this manner undermine the system's security) by putting on a show to be agreeable yet as a result dropping any information they are intended to pass on. These activities may result in defragmented systems, disengaged hubs and radically decreased system execution. The proposed methodology expects to assess the additional impact of the nearness of noxious hubs on a system execution and decide fitting measures to recognize malevolent hubs.

In applications where verification isn't basic, there is as yet a requirement for instruments whereby hubs can be guaranteed, that bundles will be conveyed to their planned goal. To address this need, this section proposes the utilization of 'respectability based' steering tables to identify and separate noxious hubs. In this research plan, every hub is checked and doled out with 'conviction scores,' in light of trust as indicated by their watched conduct and 'parcel sending history.'

Keeping up such a table at the focal hub encourages the decision confided in courses instead of the briefest ones, possibly alleviating the parcel misfortunes caused by malignant hubs, notwithstanding when confirmation isn't utilized. In the proposed approach, the cluster head alongside the accumulation of courses in its table additionally keeps up a conviction score for every hub. The bunch head checks every hub for its trust esteem and produces the key and conveys to the hub the trust esteem more than or equivalent to the limit esteem. This way the most astounding trust esteem is chosen for the transmission.

The QoS trust-based key management protocol is based on the two mechanisms whose flowchart is depicted in Figure 7.4.

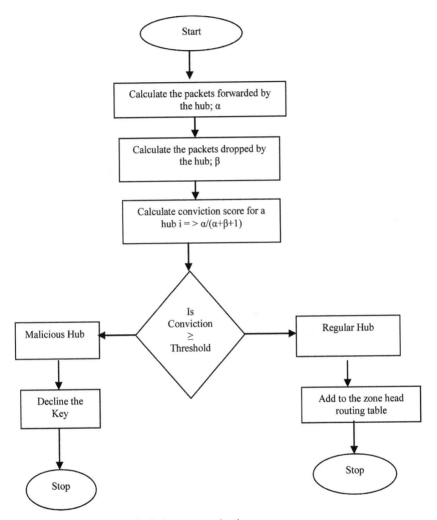

FIGURE 7.4 Flow chart depicting trust evaluation.

1. **Trust Evaluation:** It involves the trust factor of the hub which is evaluated on the basis of the number of packets forwarded and dropped. Conviction score is calculated and the path with the highest value is chosen for the packet forwarding. Trust evaluation

is performed by maintaining a conviction score, calculated with the help of α; the count of packets forwarded and β; the count of packets dropped.

$$\text{Conviction score for a hubi} = \alpha/(\alpha+\beta+1) \qquad (1)$$

where: α is the count of packets forwarded by the hubi; β is the count of packets dropped by the hubi.

After calculating the conviction value for each hub, the approach must be able to differentiate between trustworthy and non-trustworthy (malicious) hubs. Different researchers have represented trust values differently. However, they can be represented continuously in the range $(-1, +1)$; where $+1$ refers to highly trustworthy and -1 refers to non-trustworthy. This thesis takes .75 as the threshold value to determine if a hub is trustworthy.

Any hub with conviction value less than .75 is considered to be malicious. If conviction value $\geq$ .75, the hub is considered trustworthy and the generated key is passed to the hub; if conviction value < .75, the hub is marked malicious and is not included in the route. If the trust counter value falls below a trust threshold, the corresponding hub is marked as malicious, thus maintaining QoS in a network.

2. **Key Distribution:** The key distribution approach constitutes of:
   i.   Key generation using hash function;
   ii. Key management approach.

Both the operations are performed by the cluster head.

i. **Key Generation Using Hash Function:** It uses hash function to generate the keys and the cluster head distributes the keys after calculating the trust values of the hubs. A hub must have the key for receiving the packet. The key generation approach utilizes the message digest algorithm (MD5) hash algorithm for generating unique keys. MD5 is the cryptographic hash function which accepts input as message of arbitrary length and generates fixed length message digest value. The output is a 128-bit digest value. The outputs created are unique for each input and it is computationally infeasible to generate two messages with the same message digest as depicted in Figure 7.5.

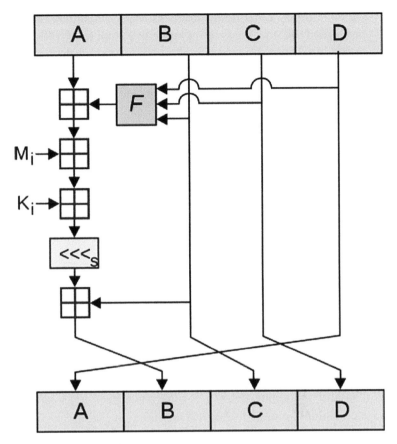

FIGURE 7.5 MD5 operation stage.

MD5 algorithm accepts any variable length message as input and processes a fixed length output contained of 128-bits. The variable length message is first appended with bits to bring the length to a multiple of 512. The 512-bit message is obtained after padding the bits to the message that is divided into sixteen 32-bit words. The computation of the final message digest output value is executed in distinct stages. Each stage processes a 512-bit data block which it breaks down into sixteen words of 32-bits each. The algorithm performs a total of 64 operations to deduce an output of 128 bits. The 64 operations are divided into four rounds. Each round is composed of 16 operations on the sixteen 32-bit words message input. Each round involves sixteen identical operations performed on the basis of a nonlinear function, followed by modular addition, and then finally by left rotating the bits. Figure 7.5 demonstrates one such operation in the first round; A, B, C, and D as shown in Figure 7.5 forms the four 32 bit words which are set to a constant value.

     refers to the addition modulo 2^{32}

     refers to the left bit rotation where s defines the number of places and varies from one operation to another.

M_i refers to the one of the sixteen 32-bit blocks.
K_i refers to the 32-bit constant value.
F refers to the non-linear function different for each round containing 16 operations.

The 128-bit output is unique for every message and is used in this proposal as the key. The key acts as a token for the hub. The hub containing the key is only considered to be safe and is included in the route to the destination.

ii. **Key Management Approach:** This approach ensures that the hub with the conviction score greater than the threshold value is only given the key as token. In route discovery, only the hubs with the keys are selected to form the path to the destination. This approach manages the elimination of malicious hubs from forming a part of the route and thus proves to be an effective mechanism in combating the attacks caused by malicious hubs.

7.9 CONCLUSIONS AND FUTURE PROSPECTS

The research study focused on this chapter displays an extensive survey of trust in informal organizations. The chapter analyzed the definitions and estimations of trust through the crystals of human science, brain science, and software engineering to maintain QoS in networks. This investigation distinguished the different features and parts of trust as: calculative, social, enthusiastic, subjective, institutional, and dispositional. The investigation likewise distinguished the different properties of trust as: reliant, dynamic, transitive, propagative, composable, customized, hilter-kilter, self-strengthening, and occasion delicate. Numerous parts of trust have been considered in various controls as the investigation of trust in informal communities is still in a beginning period. Trust models in informal organizations can be to a great extent viewed as adjustments models from different orders to interpersonal organizations. Accordingly, trust models in informal organizations are yet to incorporate most parts of trust in displaying. The trust appraisement paradigm by QoS accretion proposes calculating the conviction value for each hub, hence differentiate between trustworthy and non-trustworthy (malicious) hub. Different researchers have represented trust values differently. However, they can be represented continuously in the range (–1, +1); where +1 refers to highly trustworthy and –1 refers to non-trustworthy. The proposed approach takes.75 as the threshold value to determine if a hub is trustworthy. Any hub with conviction value less than.75 is considered to be malicious. If conviction value ≥.75, the hub is considered trustworthy and the generated key is passed to the hub; if conviction value <.75, the hub is marked malicious and is not included in the route. If the trust counter value falls below a trust threshold, the corresponding hub is marked as malicious thus maintaining QoS in a network. Hence, the significance of trust is assessed and the predecessors distinguished in this examination could be utilized. Likewise, social capital association with goal towards trust expansion ought to be assessed all the more cautiously. This would likewise help decide causalities between social capital elements, trust, and utilization. As web is an essential part of each one's life nowadays, a way is required in establishing trust in web. At the point when there exists a trust association with a long-range interpersonal communication benefit, trust, and its impact utilization could be lost. The predecessors could likewise be stretched out to different parts of web-based life, for example, web journals, which are an essential data hotspot for some individuals. The significance of the forerunners could be diverse in online life administrations as stage qualities do not impact connections but their framework. The precursors are utilized to survey whether to trust or not to confide in the administration. Human collaborations and vulnerabilities in person to person communication request trust to be the most critical factor for investigation and assessment.

KEYWORDS

- **conviction score**
- **denial of service**
- **friend-of-a-friend**
- **message-digest algorithm (MD5)**
- **quality of service**
- **trust association and management**

REFERENCES

Adali, S., Escriva, R., Goldberg, M. K., Hayvanovych, M., Magdon-Ismail, M., Szymanski, B. K., Wallace, W. A., & Williams, G., (2010). Measuring behavioral trust in social networks. In: *Proceedings of the IEEE International Conference on Intelligence and Security Informatics (ISI'10)* (pp. 150–152).

Artz, D., & Gil, Y., (2007). A survey of trust in computer science and the semantic web. *Web Semantics, 5*(2), 58–71.

Azer, M. A., El-Kassas, S. M., Hassan, A. W. F., & El-Soudani, M. S., (2008). A survey on trust and reputation schemes in ad hoc networks. In: *Proceedings of the 3rd International Conference on Availability, Reliability and Security. IEEE Computer Society, Los Alamitos, CA* (pp. 881–886).

Barka. E., et al., (2018). Union: A trust model distinguishing intentional and unintentional misbehavior in Inter-UAV communication. *Journal of Advanced Transportation, 12.* Article ID 7475357. https://doi.org/10.1155/2018/7475357 (accessed on 16 February 2020).

Barnes, J. A., (1984). Class and committees in a Norwegian island parish. *Hum. Relat., 7*(1), 39–54.

Beatty, P., Reay, I., Dick, S., & Miller, J., (2011). Consumer trust in e-commerce web sites: A metastudy. *ACM Comput. Surv., 43*(3), 1–46.

Brunie, A., (2009). Meaningful distinctions within a concept: Relational, collective, and generalized social capital. *Social Sci. Res., 8*, 2, 251–265.

Buskens, V., (1998). The social structure of trust. *Social Netw., 20*(3), 265–289.

Chen, H. C., & Chen, A. L. P., (2011). A music recommendation system based on music data grouping and user interests. In: *Proceedings of the 10th International Conference on Information and Knowledge Management (CIKM'01)* (pp. 231–238). ACM Press, New York.

Chen, L., & Li, J., (2001). Revocation of direct anonymous attestation. In: *Proceedings of the 2nd International Conference on Trusted Systems (INTRUST'10)* (pp. 128–147).

Coleman, J. S., (1988). Social capital in the creation of human capital. *Amer. J. Sociology, 94*(1), 95–120.

Coleman, J. S., (1990). *Foundations of Social Theory.* Belknap Press, Cambridge, MA.

Coulter, K. S., & Coulter, R. A., (2002). Determinants of trust in a service provider: The moderating role of length of relationship. *J. Service. Market*, *16*(1), 35–50.

Dumouchel, P., (2005). Trust as an action. *Euro. J. Sociol.*, *46*, 417–428.

Elsalamouny, E., Sassone, V., & Nielsen, M., (2010). HMM-based trust model. In: *Proceedings of the 6th International Workshop on Formal Aspects on Security and Trust (FAST'09) 2Lecture Notes in Computer Science* (Vol. 5983, pp. 21–35). Springer.

Golbeck, J. A., (2005). *Computing and Applying Trust in Web-Based Social Networks*. PhD thesis, University of Maryland at College Park, MD.

Golbeck, J., (2005). Personalizing applications through integration of inferred trust values in semantic web based social networks. In: *Proceedings of the Semantic Network Analysis Workshop at the 4th International Semantic Web Conference* 2006.

Golbeck, J., (2006). Combining provenance with trust in social networks for semantic web content filtering. In: *Proceedings of the International Conference on Provenance and Annotation of Data (IPAW'06) Lecture Notes in Computer Science Series* (Vol. 4145, pp. 101–108). Springer.

Golbeck, J., (2006). Trust on the world wide web: A survey. *Found. Trends Web Sci.*, *1*(2), 131–197.

Golbeck, J., (2007). The dynamics of web-based social networks: Membership, relationships, and change. *First Monday*, *12*, 11.

Golbeck, J., Parsia, B., & Hendler, J., (2003). Trust networks on the semantic web. In: *Proceedings of the 7th International Workshop on Cooperative Intelligent Agents* (pp. 238–249).

Grandison, T., & Sloman, M., (2000). A survey of trust in internet applications. *IEEE Comm. Surv. Tutorials*, *3*, 4.

Hoffman, K., Zage, D., & Nita-Rotaru, C., (2009). A survey of attack and defense techniques for reputation systems. *ACM Comput. Surv.*, *42*, 1–31.

Holmes, J., (1991). Trust and the appraisal process in close relationships. In: *Advances in Personal Relationships.*

Huynh, T. D., Jennings, N. R., & Shadbolt, N. R., (2006). Certified reputation: How an agent can trust a stranger. In: *Proceedings of the 5th International Joint Conference on Autonomous Agents and Multiagent Systems (AAMAS'06)* (pp. 1217–1224). New York, NY.

Jain, A. K., Tokekar, V., & Shrivastava, S., (2018). Security enhancement in MANETs using fuzzy-based trust computation against black hole attacks. In: *Information and Communication Technology, Advances in Intelligent Systems and Computing* (p. 625). Springer, Singapore.

Jhaver, R H., et al., (2018). Sensitivity analysis of an attack-pattern discovery based trusted routing scheme for mobile Ad-Hoc networks in industrial IoT. *Special Section on Security and Trusted Computing for Industrial Internet of Things, IEEE Access*. 10.1109/ACCESS.2018.2822945.

Johnson, D., & Grayson, K., (2005). Cognitive and affective trust in service relationships. *J. Bus. Res.*, *58*(4), 500–507.

Josang, A., & Golbeck, J., (2009). Challenges for robust trust and reputation systems. In: *Proceedings of the 5th International Workshop on Security and Trust Management* (Stm'09).

Josang, A., & Ismail, R., (2009). The beta reputation system. In: *Proceedings of the 15th Bled Conference on Electronic Commerce* (pp. 891–900).

Josang, A., (2001). A logic for uncertain probabilities. *Int. J. Uncert., Fuzzin. Knowl.-Based Syst.*, *9*(3), 279–311.

Josang, A., Gray, E., & Kinateder, M., (2003). Analysing topologies of transitive trust. In: *Proceedings of the 1st Workshop on Formal Aspects in Security and Trust (Fast'03)* (pp. 9–22).

Josang, A., Hayward, R., & Pope, S., (2006). Trust network analysis with subjective logic. In: *Proceedings of the 29th Australasian Computer Science Conference (Acsc'06)* (pp. 85–94). Australian Computer Society, Hobart, Australia.

Josang, A., Ismail, R., & Boyd, C., (2007). A survey of trust and reputation systems for online service provision. *Decis. Support Syst., 43*(2), 618–644.

Kerr, R., & Cohen, R., (2009). Smart cheaters do prosper: Defeating trust and reputation systems. In: *Proceedings of the 8th International Conference on Autonomous Agents and Multiagent Systems (Aamas'09)* (Vol. 2, pp. 993–1000).

Lewis, J. D., & Weigert, A., (1985). Trust as a social reality. *Social Forces, 63*(4), 967–985.

Liu, H., Lim, E. P., Lauw, H. W., Le, M. T., Sun, A., Srivastava, J., & Kim, Y. A., (2008). Predicting trusts among users of online communities: An epinions case study. In: *Proceedings of the 9th ACM Conference on Electronic Commerce (Ec'08)* (pp. 310–319). ACM Press, New York.

Malik, Z., & Bouguettaya, A., (2009). Reputation bootstrapping for trust establishment among web services. *IEEE Internet Comput., 13*, 40–47.

Malik, Z., Akbar, I., & Bouguettaya, A., (2009). Web services reputation assessment using a hidden Markov model. In: *Proceedings of the 7th International Joint Conference on Service-Oriented Computing (Icsocservicewave'09)* (pp. 576–591).

Marsh, S. P., (1994). *Formalizing Trust as a Computational Concept.* PhD Thesis University of Stirling.

Mcleod, A., & Pippin, S., (2009). Security and privacy trust in e-government: Understanding system a relationship trust antecedents. In: *Proceedings of the 42nd Hawaii International Conference On System Sciences (Hicss'09)* (pp. 1–10).

Mollering, G., (2002). The nature of trust: From Geog Simmel to a theory of expectation. *Interpretation and Suspension, Sociol., 35*, 403–420.

Momani, M., & Challa, S., (2010). Survey of trust models in different network domains. *Int. J. Ad. Hoc. Sensor. Ubiq. Comput., 1*(3), 1–19.

Moreland, D., Nepal, S., Hwang, H., & Zic, J., (2010). A snapshot of trusted personal devices applicable to transaction processing. *Personal Ubiq. Comput., 14*(4), 347–361.

Mui, L., (2003). Computational models of trust and reputation: Agents, evolutionary games, and social networks. PhD Thesis.

Mui, L., Mohtashemi, M., & Halberstadt, A., (2002). A computational model of trust and reputation. In: *Proceedings of the 35th Hawaii International Conference on System Sciences (Hicss'02), IEEE Computer Society, Los Alamitos, Ca* (pp. 188–196).

Richardson, M., Agrawal, R., & Domingos, P., (2002). Trust management for the semantic web. In: *Proceedings of the 2nd International Semantic Web Conference* (pp. 351–368).

Rotter, J. B., (1967). A new scale for the measurement of interpersonal trust. *J. Personality, ACM Computing Surveys, 45*(4), Article 47.

Ruohomaa, S., & Kutvonen, L., (2005). Trust management survey. In: *Proceedings of the 3rd International Conference on Trust Management (Itrust'05)* (pp. 77–92). Springer.

Ruohomaa, S., Kutvonen, L., & Koutrouli, E., (2007). Reputation management survey. In: *Proceedings of the 2nd International Conference on Availability, Reliability and Security (Ares'07)* (pp. 103–111). IEEE Computer Society.

Sabater, J., & Sierra, C., (2002). Reputation and social network analysis in multi-agent systems. In: *Proceedings of the 1ˢᵗ International Joint Conference on Autonomous Agents and Multiagent Systems (Aamas'02)* (pp. 475–482). ACM Press, New York.

Sabater, J., & Sierra, C., (2005). Review on computational trust and reputation models. *Artif. Intell. Rev.*, *24*, 33–60.

Sabater, J., (2002). *Trust and Reputation for Agent Societies*. PhD Thesis, Autonomous University of Barcelona, Spain.

Scully, M., & Preuss, G., (1996). Two faces of trust: The roles of calculative and relational trust in work transformation. *Tech. Rep. Working Paper No.*, 3923–3996.

Seshadri, A., Perrig, A., Van Doorn, L., & Khosla, P. K., (2004). Swatt: Software-based attestation for embedded devices. In: *Proceedings of the IEEE Symposium on Security and Privacy* (pp. 272–282).

Singh, S., & Bawa, S., (2007). Privacy, trust and policy-based authorization framework for services in distributed environments. *Int. J. Comput. Sci.*, *2*(2), 85–92.

Song, S., Hwang, K., Zhou, R., & Kwok, Y. K., (2005). Trusted P2p transactions with fuzzy reputation aggregation. *IEEE Internet Comput.*, *9*(6), 24–34.

Song, W., Phoha, V. V., & Xu, X., (2004). The hmm-based model for evaluating recommender's reputation. In: *Proceedings of the IEEE International Conference on E-Commerce Technology for Dynamic E-Business* (pp. 209–215). IEEE.

Suh, B., & Han, I., (2002). Effect of trust on customer acceptance of internet banking. *Electron. Commerce Res. Appl.*, *1*(3/4), 247–263.

Suryanarayana, G. T. R., (2004). A survey of trust management and resource discovery technologies. In: *Peerto-Peer Applications, Tech. Rep. Uci-Isr-04-6.* Institute for Software Research, University of California, Irvine.

Taylor, R., (2000). Marketing strategies: Gaining a competitive advantage through the use of emotion. *Competitive. Rev.*, *10*, 146–152.

Vianna, S., & Priya, V. S., (2018). Trust-based approach to overcome black hole attack in MANET. *International Journal of Pure and Applied Mathematics*, *118*(22), 1763–1769. ISSN: 1314–3395 (on-line version) url: http://acadpubl.eu/hub (accessed on 16 February 2020).

Waldman, M., Rubin, A. D., & Cranor, L. F., (2013). Publius: A robust, tamper-evident, censorship resistant web publishing system. In: *Proceedings of the 9ᵗʰ Usenix Security Symposium* (Vol. 45, No. 4). ACM Computing Surveys, Article 47, Publication Date: August.47:32 W 2000.

Wang, T., & Lu, Y., (2010). Determinants of trust in e-government. In: *Proceedings of the International Conference on Computational Intelligence and Software Engineering (Cise'10)* (pp. 1–4).

Wang, Y., & Vassileva, J., (2007). A review on trust and reputation for web service selection. In: *Proceedings of the 27ᵗʰ International Conference on Distributed Computing Systems Workshops (Icdcsw'07)* (pp. 25–32). IEEE Computer Society.

Wang, Y., Hori, Y., & Sakurai, K., (2007). Economic-inspired truthful reputation feedback mechanism in P2p networks. In: *Proceedings of the 11ᵗʰ IEEE International Workshop on Future Trends of Distributed Computing Systems* (pp. 80–88).

Xiong, L., & Liu, L., (2003). A reputation-based trust model for peer-to-peer ecommerce communities. In: *Proceedings of the 4ᵗʰ ACM Conference on Electronic Commerce (Ec'03)* (pp. 228–229). ACM Press, New York.

Xiong, L., & Liu, L., (2004). Peer trust: Supporting reputation-based trust for peer-to-peer electronic communities. *IEEE Trans. Knowl. Data Engin.*, *16*(7), 843–857.

Yao, J., Chen, S., Nepal, S., Levy, D., & Zic, J., (2010). Truststore: Making Amazon S3 trustworthy with services composition. In: *Proceedings of the 10th IEEE/ACM International Conference On Cluster, Cloud and Grid Computing (Ccgrid'10)* (pp. 600–605). IEEE Computer Society, Los Alamitos, Ca.

Yu, B., & Singh, M. P., (2000a). A social mechanism for reputation management in electronic communities. In: *Proceedings of the 4th International Workshop on Cooperative Information Agents (Cia'00)* (pp. 154–165). Springer.

Yu, B., & Singh, M. P., (2002b). An evidential model of distributed reputation management. In: *Proceedings of the 1st International Joint Conference on Autonomous Agents and Multi-Agent Systems (Aamas'02)* (pp. 294–301). ACM Press, New York.

Yu, B., Singh, M. P., & Sycara, K., (2004). Developing trust in large-scale peer-to-peer systems. In: *Proceedings of the 1st IEEE Symposium on Multi-Agent Security and Survivability* (pp. 1–10). IEEE Computer Society.

Zhou, R., & Hwang, K., (2007). *Power Trust: A Robust and Scalable Reputation System for Trusted Peer-to-Peer Computing* (Vol. 18, No. 4, pp. 460–473). IEEE Trans. Parallel Distrib. Syst.

CHAPTER 8

Collaborative Filtering in Recommender Systems: Technicalities, Challenges, Applications, and Research Trends

PRADEEP KUMAR SINGH, PIJUSH KANTI DUTTA PRAMANIK, and PRASENJIT CHOUDHURY

Department of Computer Science and Engineering,
National Institute of Technology Durgapur, India
E-mail: pijushjld@yahoo.co.in (P. K. D. Pramanik)

ABSTRACT

The rapid development and extensive use of recommender systems (RSs) have changed the face of online service experience. The enormous data generated and the complexity involved in analyzing these data for an effective recommendation has attracted researchers from different domains, especially data analytics. In this direction, collaborative filtering (CF) has been the most widely considered approach. The objective of this chapter is to represent a comprehensive study of the CF. The chapter is written in a tutorial fashion so that it can be followed by the readers who are the beginners in this field or unfamiliar with the RS. Different aspects of CF such as classifications, approaches, data extraction methods, similarity metrics, prediction approaches, and performance metrics are studied meticulously. The application of CF in different domains is reviewed. More than 100 research articles are surveyed and categorized according to the application domain of CF they have covered. The challenges involved in the successful adoption of the CF are validly examined. In addition to a brief survey on CF, a systematic survey, considering 277 related papers, on current research trends (2011–2017) on CF is presented. A special discussion of future directions of CF is also stated.

8.1 INTRODUCTION

The recommender system (RS) has become the backbone of e-commerce. In addition to the basic searching facility, every e-commerce portal is opting for RS as an integral part of it. As the e-commerce market is continuously growing, more products and services are made available for online purchasing. Among this sea of online products and services, customers find it very difficult to find the appropriate item for themselves. The e-commerce vendors have come up with the solution for helping the customer to find the appropriate item by recommending the item to the customer which he/she might like or desire. The technical scheme that enables the recommendation process is termed as a RS. RS attempts to predict the items that prospective online buyers may prefer and recommend these anticipated items. Unlike the search tools where people ferret out the online products, recommendation engines aim to consciously catch the attention of the users to the likable products. The overall objective is to bail out users form explicit and tiresome searching and to improve the online shopping experience. The success of e-business largely depends on the intelligence of the algorithm used for a product recommendation. Hence, in the age of digital marketing, it is crucial for online stores to adopt intelligent recommendation techniques in order to sustain in the market competition. Companies like Flipkart, Amazon, eBay, Netflix, MovieLens, IMDb, etc. use RS extensively and innovatively as a core part of their business innovation and exploration.

RS assesses the preference and choice of users by tracking and analyzing their buying and browsing habits and history. The tool used for this purpose is generally known as the filtering approach. Filtering Approach is a method which makes the selective presentation among an array of available commodities using various filtering parameters which makes the filtered products more favorable to the recipient. There are several filtering approaches of an RS in the literature such as: (i) content-based (CB), (ii) CF, (iii) hybrid filtering, (HB), (iv) knowledge-based (KB), and (v) context-aware (CA) as shown in Figure 8.1. CF is more popular filtering approach among these over the past few years (Burke, 2002). CF works on the fact of comparison of user activities, purchases, ratings, preferences, and using this data for comparison and subsequent analysis. The customers prefer products that have been liked or given higher preference by people with a similar taste (Deshpande and Karypis, 2004). Hence, the CF is the most important in this regard.

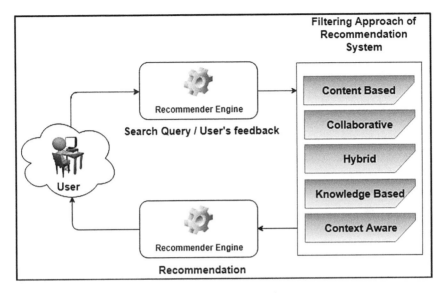

FIGURE 8.1 A general framework for the recommender system.

CF attempts to guess the target user's interest by assessing the top-n similar users' interests on the basis of the assumption that if two persons' choice matches for certain things, it is highly probable that their choices will match for other things as well. The CF suggests that the ratings given by similar users tend to be substantively similar and similar items also tend to receive similar ratings. The CF algorithms exploit this assumption for the recommendation and actually use the similarity value to predict user preferences. Similarity allows the recommender engines to find the user purchase patterns as well as allowing them to understand how those rating patterns are similar to other users. All rating information is stored into memory for prediction in making of the top-n list for recommendation. Similar users or items have a major contribution in the prediction phase of CF-based RS. The top-n list of the recommended items affected if similarity provides the wrong result. CF uses two approaches for considering similarity (Singh, Pramanik, and Choudhury, 2018):

    i. **User Similarity-Based Approach (USBA):** Tries to predict the rating based on rating information collected from similar users; and

    ii. **Item Similarity-Based Approach (ISBA):** Uses the same idea as USBA but, it uses item similarity instead of user similarity.

After computation of the similarity value of users' and items,' prediction approaches are used to predict the ratings of a target item for a target user. Furthermore, CF-based RS generates a list of top-n items and recommend to the target user. The top-n list of the recommended items affected if similarity provides the wrong result.

The structure of the remaining chapter is as follows. Several application domains of RS are stated in Section 8.2. More than 100 papers are surveyed and categorized according to the application domain they address and the filtering approach they used. Section 8.3 mentions different CF approaches. The working principle of neighborhood-based CF is explained in Section 8.4. Section 8.5 introduces the data extraction methods used in CF. Section 8.6 explains the similarity metrics used in CF algorithms while a comparative study on different similarity metrics has been presented in Section 8.7. Sections 8.8 and 8.9 discuss the prediction approaches and performance metrics used in CF-based RSs, respectively. The challenges in CF-based RSs, as well as the security and trust attacks, are discussed meticulously in Section 8.10. Section 8.11 mentions some of the notable works on CF. Section 8.12 presents the research trends in CF-based RS. 277 related papers are studied for this purpose. The future scope of CF-based Rs is discussed briefly in Section 8.13. And finally, the conclusion of the paper is presented in Section 8.14.

8.2 APPLICATIONS OF RECOMMENDER SYSTEMS (RSS)

RS has found many application domains. Below a few of them mentioned:

1. **E-Government:** It is the medium by which the government makes use of the internet and computers to deliver services to the citizens. It is the most effective modern method which helps the government to connect people across the country.

2. **E-Library and E-Learning:** It is the medium by which the system of education is provided to individuals completely over the internet with the help of electronic devices. It is a formal way of delivering education through electronic resources.

3. **E-Tourism:** It is the digital process which is implemented to achieve the strategy of e-commerce in tourism. It also helps to keep the client connected with the travel partners. E-tourism leads to an excellent medium of marketing and promotions of a company.

4. **E-Resource:** It refers to any resource or collection preserved in electronic format. This type of resource requires an electronic device to access the information. Since the resource is available in the electronic format, huge sets of data can be available for access.

5. **E-Commerce:** Any commercial transactions, exchange, or transfer of data which is carried on via the internet is termed as electronic commerce or e-commerce. It is the fastest method of conducting business in the modern world and thus leads to the digitization of society.

The performance of these applications can be improved using memory-based CF. People can easily provide their opinion about the services of these applications and due to this; they can be received more personalized, diverse, novel, and accurate recommendations. Table 8.1 lists different application domains of RS. It also mentions the notable research works towards these domains and also the filtering approaches used in those works.

TABLE 8.1 Application Domains of Recommender Systems, Notable Works Towards That Domain, and Filtering Approach Used

Application Domain	Filtering Approach	Recommender System
E-government	Knowledge-based	Meo, Quattrone, and Ursino, 2008; Teran and Meier, 2010; Esteban et al., 2014; Cornelis et al., 2007
	Collaborative	Guo and Lu, 2007
	Collaborative, Hybrid, Knowledge-based	Wu, Zhang, and Lu, 2015; Lu et al., 2010
E-library and e-learning	Content-based, Collaborative, Hybrid	Balabanović and Shoham, 1997; Renda and Straccia, 2005
	Hybrid, Knowledge-based	Porcel, López-Herrera, and Herrera-Viedma, 2009; Porcel, Herrera-Viedma, and Moreno, 2009; Porcel and Herrera-Viedma, 2010; Serrano-Guerrero et al., 2011; Cobos et al., 2013
	Knowledge-based, Content-based	Zaíane, 2002; Chen, Duh, and Liu, 2004; Chen and Duh, 2008; Capuano et al., 2014; Farzan and Brusilovsky, 2006; Santos et al., 2014; Lu, 2004; Biletskiy et al., 2009
E-tourism	Knowledge-based	Burke, Hammond, and Young, 1996; Fesenmaier et al., 2003; García-Crespo et al., 2011

TABLE 8.1   *(Continued)*

Application Domain	Filtering Approach	Recommender System
	Knowledge-based, Collaborative, Context-aware, Hybrid	Avesani, Massa, and Tiella, 2005; Martínez, Rodríguez, and Espinilla, 2009; Ruotsalo et al., 2013; García-Crespo et al., 2009; Console et al., 2003; Moreno et al., 2013
	Content-based, Collaborative, Hybrid, Demographic	Schiaffino and Amandi, 2009; Luz et al., 2013; Baraglia et al., 2012
	Context-aware	Tung and Soo, 2004; Pashtan et al., 2003; Rikitianskii, Harvey, and Crestani, 2014; Xing et al., 2013
	Collaborative, Context-aware	Yanga and Hwang, 2013
E-resource	Content-based	Jinni, 2017; Rotten Tomatoes, 2017; IMDb, 2017; Asnicar and Tasso, 1997; ACRnews, 2017; Chesnevar and Maguitman, 2004; Park, 2013
	Collaborative	Ali and Van Stam, 2004; Konstan et al., 1997; FoxTrit, 2017; Miller, Konstan, and Riedl, 2004; Hauver and French, 2001; Marcel et al., 2003; Lee, Cho, and Kim, 2010; TASTEKiD, 2017; nanoCROWD, 2017; Movielens, 2017
	Context-aware, Collaborative	Braunhofer, Kaminskas, and Ricci, 2013; Baltrunas et al., 2012; Levandoski et al., 2012; Natarajan, Shin, and Dhillon, 2013; Oh et al., 2014
	Collaborative, Knowledge-based	Zhang, Zhou, and Zhang, 2011; Hayes and Cunningham, 2001; Sánchez et al., 2011; Boutet et al., 2013
	Content-based, Collaborative, Hybrid	Smyth and Cotter, 2000; Blanco-Fernández et al., 2006; Salter and Antonopoulos, 2006; Melville, Mooney, and Nagarajan, 2002; Domingues et al., 2013; Christou, Amolochitis, and Tan, 2016; Parra, Brusilovsky, and Trattner, 2014; Amolochitis, Christou, and Tan, 2014
	Knowledge-based, Content-based	Jäschke et al., 2007; Hotho et al., 2006; Celma and Serra, 2008; Bjelica, 2010; Moukas and Maes, 1998; Billsus and Pazzani, 2000; Nguyen, Lu, and Lu, 2014; Martín-Vicente et al., 2012; Zhang et al., 2012

TABLE 8.1 *(Continued)*

Application Domain	Filtering Approach	Recommender System
E-commerce	Knowledge-based, Demographic	Garfinkel et al., 2006; Mccarthy et al., 2004; Cao and Li, 2007; Hu et al., 2012; Zhao et al., 2016; Zhao et al., 2014;
	Knowledge-based, Content-based	Burke, 1999; Nanopoulos et al., 2010; Zhang et al., 2013; Yin et al., 2014
	Collaborative, Hybrid	Pratikshashiv, 2015; Lawrence et al., 2001; Chen and Pu, 2012; Walter et al., 2012; Liu and Karger, 2015

8.3   COLLABORATIVE FILTERING (CF) APPROACHES

The CF technique can be classified into two categories as shown in Figure 8.2 (Su and Khoshgoftaar, 2009):

i.   **Model-Based CF:** It uses some algorithms of machine learning (ML) like Bayesian network clustering and rule-based approaches which builds a model on user-item rating dataset and then recommends items to the user.

ii   **Neighborhood/Memory-Based CF:** Similarity and prediction computation are the two major steps used in this category of CF.

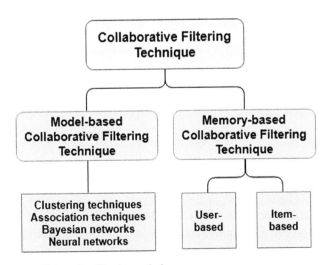

FIGURE 8.2   Collaborative filtering techniques.

8.4 WORKING PRINCIPLE OF NEIGHBORHOOD-BASED COLLABORATIVE FILTERING (CF)

Figure 8.3 shows the conceptual framework of neighborhood-based CF (Yang et al., 2016). Neighborhood CF defines the closest neighbors using the following two algorithms:

- **User-Based CF Algorithm:** User similarity metric is used to find the nearest neighbors. The rating value of these neighbors and their similarity values are utilized in the prediction of unrated items of users for the formation of the Top-n list in the recommendation.
- **Item-Based CF Algorithm:** In the item-based CF algorithm, the nearest neighbors are determined using the similarity values of items, and these similarity values and rating values of these neighbors are used in the formation of the recommendation list to the user.

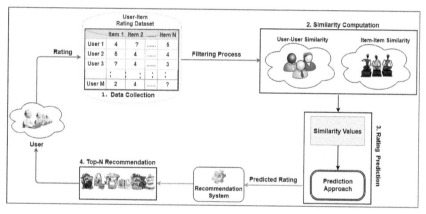

FIGURE 8.3   A conceptual framework for neighborhood-based collaborative filtering.

Table 8.2 shows the descriptions of the notations used in this chapter.

8.5 DATA EXTRACTION METHODS USED IN COLLABORATIVE FILTERING (CF)

CF uses ratings in the recommendation process. Two types of ratings have been used in CF for a recommendation-explicit rating and implicit rating (Li et al., 2018).

TABLE 8.2 Notations and Their Descriptions

Notation	Description
Sim(i,j)	Similarity between two items i and j
$R_{u,i}$	Rating value of user u on item i
$\bar{R}_u$	Average or mean rating value of user u
$|U_{ij}|$	Number of ratings of user u on both items i and j
$\widehat{r_{ui}}$	Predicted rating value of user u on item i
$\bar{r}_i$	Average or mean rating value of item i

1. **Explicit:** These ratings are the specific rating that a user gives to a product (for example, a user rates a book 3 on a scale of 1 to 5). These explicit ratings are directly used in the extractions of users' interest for future recommendation. The disadvantage of explicit data is that it makes user responsible for data collection and future rating prediction who hardly takes interest to give a rating on a particular item.

2. **Implicit:** These ratings are collected by logging the user's data generated while browsing the website. Implicit data are easier to collect as it does not put any pressure on the user to rate the products on the site. However, dealing with an implicit rating is very complicated as it is hard to find the users' preferences from these collected users' browsing data. Using these collected ratings (explicit or implicit); RSs predict the unknown ratings of the user based on different similarity metrics and these predicted ratings used in the recommendation process.

8.6 SIMILARITY METRICS USED IN COLLABORATIVE FILTERING (CF) ALGORITHMS

There are various similarity metrics used in the CF to find the nearest neighbors and similarity values (Sarwar et al., 2001; Bilge and Kaleli, 2014; Bobadilla et al., 2012). The metrics used in the item-based CF are:

1. **Cosine Similarity (CS):** The function of cosine distance finds similarity between two samples by studying the cosine of the angle between them to quantify the similarity. The similarity values are in the range [1, −1], where 1 shows the maximum similarity and −1

depicts no similarity. CS between two items i and j, is calculated using:

$$sim(i, j) = cos(i, j) = \frac{i.j}{||i||^2 * ||j||^2}$$

Here, i and j identifies the dot-product between two items.

2. **Adjusted Cosine Similarity (ACS):** It is similar to cosine distance, also caters to the individual user's rating. To achieve this, it subtracts the average user rating from the individual ratings to get uniformity. It is computed by:

$$sim(i, j) = \frac{\sum_{u \in U} \left(R_{u,i} - \bar{R}_u \right) \left(R_{u,j} - \bar{R}_u \right)}{\sqrt[2]{\sum_{u \in U} \left(R_{u,i} - \bar{R}_u \right)^2} \sqrt[2]{\sum_{u \in U} \left(R_{u,j} - \bar{R}_u \right)^2}}$$

Here, $R_{u,i}$ and $R_{u,j}$ are the rating value of user u on two items i and j, respectively. $\bar{R}_{u,}$ shows the average rating value of user u.

3. **Pearson Correlation (PC):** It is the most popular Similarity Metric and is widely used in various experiments. The similarity in it is represented between $[1, -1]$, where 1 shows the maximum similarity and -1 depicts no similarity. Similarity using PC, in Item-based CF algorithm is:

$$sim(i, j) = \frac{\sum_{u \in U} \left(R_{u,i} - \bar{R}_i \right) \left(R_{u,j} - \bar{R}_j \right)}{\sqrt[2]{\sum_{u \in U} \left(R_{u,i} - \bar{R}_i \right)^2} \sqrt[2]{\sum_{u \in U} \left(R_{u,j} - \bar{R}_j \right)^2}}$$

Here, $\bar{R}_i$ and $\bar{R}_j$ are the mean rating value of two items i and j, respectively.

4. **Jaccard Similarity (JS):** It considers only all the common ratings between items in spite of the absolute rating value of items. It is calculated by [1]:

$$sim(i, j) = \frac{|R_i \cap R_j|}{|R_i \cup R_i|}$$

5. **Spearman Correlation (SC):** It is calculated just like PC, but it uses the respective rank of the actual rating value. The equation of calculating similarity value by SC is as follows:

$$sim(i, j) = \frac{\sum_{u \in U} \left(k_{u,i} - \bar{k}_i \right) \left(k_{u,j} - \bar{k}_j \right)}{\sqrt[2]{\sum_{u \in U} \left(k_{u,i} - \bar{k}_i \right)^2} \sqrt[2]{\sum_{u \in U} \left(k_{u,j} - \bar{k}_j \right)^2}}$$

Here, $k_{u,i}$ and $k_{u,j}$ show the respective rank of items i and j of rating value of user u. $\bar{k_i}$ and $\bar{k_j}$ denote the average rank of items i and j respectively.

6. **Euclidean Distance (ED):** The Euclidian distance uses the under-root of the squared sum of the difference between individual ratings of the two samples whose similarity we want to find. The distance gives an insight into how different the rating patterns are:

$$sim(i, j) = \sqrt[2]{\frac{\sum_{u \in U_{i,j}} (r_{i,u} - r_{j,u})^2}{|U_{ij}|}}$$

7. **Manhattan Distance (MD):** The equation to find similarity using MD is given below.

$$sim(i, j) = \frac{\sum_{u \in U_{i,j}} (r_{i,u} - r_{j,u})^1}{|U_{ij}|}$$

8. **Mean Squared Distance (MSD):** It is similar to ED only difference is that the whole Euclidian distance is squared, thus removing under-root from the mathematics thus making calculations easier. The equation of MSD for calculating the similarity value is shown by:

$$sim(i, j) = \frac{\sum_{u \in U_{i,j}} (r_{i,u} - r_{j,u})^2}{|U_{ij}|}$$

8.7 COMPARISON OF DIFFERENT SIMILARITY METRICS

Purpose of the RS is to provide optimized and personalized products recommendation to the users. RS has various options to choose similarity metrics (in literature), which gives various lists of top-n recommendation items.

Table 8.3 illustrates the list of top-10 similar movies of target movie id 1, using the traditional similarity measures. It can be observed that every similarity measure has a different top-10 movies list. Hence, there is a need for a comparative study on similarity metrics to enhance the accuracy of CF. On the basis of a comparative study of similarity measures; we can improve the accuracy of RS because each similarity measures have some limitations. For constructing Table 8.3, we collect the MovieLens dataset, i.e., ml–20 m. The filtering criteria have been applied to minimize the sparsity. These filtering criteria are:

i.   Select the users who provide ratings to a minimum of 100 numbers of movies.

ii. Select the movies which are received to a minimum of 1000 number of ratings.

TABLE 8.3   List of Top-10 Similar Movies of Target Movie id 1, Using the Traditional Similarity Measures

Similarity Metric	Top-10 Similar Movies									
Pearson Correlation	926	1272	1276	623	730	869	215	95	999	1301
Cosine Distance	1276	1192	1027	956	1079	949	352	1088	215	915
Adjusted Cosine Distance	1276	352	1027	1079	926	1088	580	401	15	729
Mean Squared Distance	1276	352	1088	926	1079	729	1027	1142	1239	354
Euclidean Distance	1276	352	1088	926	1079	729	1027	1142	1239	354
Manhattan Distance	1276	29	15	1088	1079	352	85	1239	1182	119
Spearman Correlation	1276	352	1027	926	1079	869	1088	915	949	1142

8.8   PREDICTION APPROACHES

Different prediction approaches have been utilized in the prediction phase of CF-based RS (Sarwar et al., 2001; Wu et al., 2013; Herlocker et al., 1999). These methods for item-based CF are:

1.   **Mean Centering (MC):** In this approach, the mean of the target item's rating is added with the weighted average (WA) of subtraction between all available ratings of top-n similar items with their respective mean is done, using as weights the correlation values computed by the similarity measures. The equation of the MC approach to predict the rating as given below:

$$\widehat{r_{ui}} = \overline{r_i} + \frac{\sum_{j \in N_u(i)} sim(i,j)(r_{ju} - \overline{r_j})}{\sum_{j \in N_u(i)} |sim(i,j)|}$$

2.   **Weighted Average (WA):** To predict the rating for a target item, a WA of all available ratings of top-n similar items is calculated using weights as the correlation values computed by the similarity measures. The equation to predict rating using WA is:

$$\widehat{r_{ui}} = \frac{\sum_{j \in N_u(i)} sim(i,j) r_{ju}}{\sum_{j \in N_u(i)} |sim(i,j)|}$$

3. **Z Score (ZS):** Using the standard deviation of rating of the item in MC, the equation of Z-score for item-based CF is as follows:

$$\widehat{r_{ui}} = \overline{r_i} + \sigma_i \frac{\sum_{j \in N_u(i)} sim(i,j)(r_{ju} - \overline{r_j}) / \sigma_i}{\sum_{j \in N_u(i)} |sim(i,j)|}$$

Here, σ_i represents the standard deviation of the rating value of item i.

These prediction approaches have some limitations in the sparse dataset. Hence, for the more personalized and accurate recommendation, there is a need for a comparative study of prediction approaches in CF. By mutually exchanging i and j with u and v respectively, we can get the computational equation of SMs and PAs in user-based CF.

8.9 PERFORMANCE METRICS

Various performance metrics have been used in the literature of CF-based RSs (Singh, Pramanik, and Choudhury, 2018; Samundeeswary and Krishnamurthy, 2017; Zuva and Zuva, 2017; Pampın, Jerbi, and O'Mahony, 2015):

1. **Mean Absolute Error (MAE):** It is the amount of error in the rating prediction. The equation for calculating MAE is:

$$MAE = \frac{\sum_{i=1}^{N} |p_i - \hat{q}_i|}{N}$$

Here, $<p_i$ and $\hat{q}_i>$ denote each original ratings-predicted ratings pair and, N shows the total number pairs that represent original and predicted ratings pair.

2. **Root Mean Square Error (RMSE):** After some modification in the equation of MAE, we get the equation of RMSE as follows:

$$RMSE = \sqrt[2]{\frac{\sum_{i=1}^{N} (p_i - \hat{q}_i)^2}{N}}$$

3. **Coverage:** Item coverage is the percentage of items included in the recommendation list over the number of potential items:

$$Coverage_{item} = \frac{n}{N} * 100$$

User coverage is the percentage of users for whom the recommender was able to generate a recommendation list over the number of potential users.

$$Coverage_{user} = \frac{u}{U} * 100$$

Catalog coverage is the percentage of recommended user-item pairs over the total number of potential pairs. The number of recommended user-item pairs can be represented by the length of the recommended lists L.

$$Coverage_{user} = \frac{length(L)}{N * U} * 100$$

And finally, user interaction coverage is the percentage of rated predictions over the total number of ratings. Here, n, and u represent the number of items in the recommendation list and the number of users involved in the generation of this recommendation list respectively. N and U denote the number of potential items and the number of potential users, whereas L shows the number of user-item in the recommendation list.

4. **Diversity:** It measures how dissimilar recommended items are for a user. This similarity is often determined using the item's content (e.g., movie genres) but can also be determined using how similar items are rated.

$$Diversity = \frac{2}{N(N-1)} \sum_{i_j \in L(u)} (1 - sim(i_j, i_k))$$

Here, $sim(i_j, i_k)$ denotes the similarity between item j and item k.

5. **Serendipity:** It is the measure of how surprising the successful or relevant recommendations are. The probability of a recommendation is simply a function of its overall rank over *n* items:

$$P_i = \frac{n - rank_i}{n-1}$$

Here, P_i represents the probability of item i for recommendation and $rank_i$ shows the rank of item i over n items. The equation of findings unexpected recommendation is:

$$UNEXP = \frac{RS}{PM}$$

Here, PM denotes the set of recommendations generated by a primitive prediction model, and RS shows the generated recommendations. UNEXP consists that list which does not belong to RS. We define serendipity as follows:

$$Serendipity = \frac{\sum_{i=1}^{N} u(RS_i)}{N}$$

6. **Novelty:** It can be defined as:

$$Novelty = \sum_{i \in L} \frac{log_2 P_i}{n}$$

Higher novelty values represent that less popular items are being recommended, thus less well-known items are likely being surfaced for users.

The equations of computing precision, recall, F-measure, and accuracy are as follows:

1. **Precision:** It can be calculated by the fraction of the recommended items that are actually relevant to the target user.

$$Precision = \frac{t_p}{t_p + f_p}$$

2. **Recall:** It consists of the relevant items that are part of the set of recommended items. Hence, the equation of calculating recall becomes:

$$Recall = \frac{t_p}{t_p + f_n}$$

3. **F-Measure:** Precision and Recall values have been used to compute the F-measure, and the equation is:

$$F\text{-}measure = 2 * \frac{Precision * Recall}{Precision + Recall}$$

4. **Accuracy:** It shows how close a predicted rating is to the actual rating. The equation of computing accuracy as follows:

$$Accuracy = \frac{t_p + t_n}{t_p + t_n + f_p + f_n}$$

Here, t_p, f_p, t_n, and f_n denote the true positive, false positive, true negative, and false negative respectively.

8.10 CHALLENGES IN COLLABORATIVE FILTERING (CF)

8.10.1 NEW USER PROBLEM

When users newly register to an RS, they do not have any ratings in their profile ratings denote the taste or preferences of the users. In the absence of a user, since CF is based on user preferences, it is unable to recommend many of the items (Lakshmi and Lakshmi, 2014). Even when users have scanty profiles with very few ratings, CF fails to render a reliable, personalized recommendation to these users. To overcome this problem, an RS used demographic features of the user from the user's profile for the recommendation. But it also has some issues that two users, having the same profile, may not have the same intent towards a particular item.

8.10.2 NEW ITEM PROBLEM

The new item is an additional issue in cold start problems which is based on the items, recurrently added to the list (Lakshmi and Lakshmi, 2014). Firstly, the items are rated then only they can be recommended to users.

8.10.3 SPARSITY PROBLEM

It takes place when the user has used some particular product but didn't bother to rate it, and another possibility can be that the user was completely unfamiliar with the product, so he didn't rate it (Lakshmi and Lakshmi, 2014). To run over this problem one approach of RS is a clustering method. Clustering method refines the data according to the preference of the user, and by doing so; it makes it easy for recommending items. But again, some issues have to be resolved in the case of multi-level clustering.

8.10.4 SCALABILITY PROBLEM

CF works on the database that contains user-item rating, and it has some scalability issue for the users and items set in large numbers. For large item set, the complexity of CF algorithms will be too large. High scalability of CF system is required as many of the systems need to respond immediately to fulfill online requirements which make recommendations for all users based on their purchase and rating history (Alloway, 2018; Poonam, Goudar, and Sunita, 2015).

8.10.5 SYNONYMY PROBLEM

Another problem with the CF approach is the synonymy problem (Xiaoyuan and Taghi, 2009). Most of the CF algorithms are unable to find similar items with various names (synonyms). Due to this, some association problem occurs, for e.g., "kids' movie" and "kids' film" is basically the same items to be searched, but according to memory-based CF, there is no match between above two terms to compute similarity. The next problem is abbreviations which are used a lot nowadays. Sometimes users are shown different results when they search for particular data, inserting abbreviations. Here the work should include these shortened words and categories them in the same list as per their full forms. Then there come the issues which are caused due to symbols or smileys. Some users prefer smiles to give a review of some products. For example, if a user wants to say that, he liked a product; he will simply give a smiley or a thumb up and used, or thumb down for dislike. So, such symbols should also be evaluated because some sites like Amazon do not hold any importance for smileys. Such sites rather ask to write a review in a minimum of 20 words (BBC, 2018). With this, there also comes the problem of reviewing the product in different regional languages. Different users want to give a review to the product in their respective languages for e.g., Hindi ("bahutachha"), Bengali ("khubbhalo"), English ("very nice"), etc. These different languages give out the same meaning that the product is good. But, if only one language will be considered then the reviews of other users, will lack its importance. For the betterment of the RS, it is also very important to take all these issues under examination.

8.10.6 *LONG TAIL (LT) PROBLEM*

In addition to the above-mentioned problems, one major problem also will arise, namely long tail (LT) problem. This section will discuss: What LT problem is and how to deal with it?

RSs basically use the past records of the user and then it anticipates the possible future likes and dislikes of the user and recommends accordingly. A better RS would propose fewer common options to draw the user's interest. It would not recommend similar kind of items repeatedly. Diversity is related to this aspect. This aspect implies the need for recommending diverse items to the user and how different the item is with respect to each other. But the RS lacks to co-operate with this aspect which leads to LT problem (Lei, 2013). The user will be deprived of many other necessary items just because he did not rate those items or because he did not have any access to those items. This generally leads to an LT problem. LT problem is when many items remain unrated or low rated. To deal with this problem, one idea is to rank the items in different ways. There is a need for segregating the ratings of the users and then rank it. Apart from the highest-rated items, there is also a need for recommending low rated items. The low rated items do not hold any importance. Researchers face the problem in the filtering of important low rated items. There is also a need to rank the items according to the purchase history and then recommend the lowest purchased item. However, it is not important that items that are more prestigious should necessarily be at the top of the list. We can see this aspect in the case of recommending books.

The LT problem can be reduced in an RS by considering (i) accuracy, (ii) similarity, (iii) diversity, and (iv) LT (Oscar, 2010; Daniel and Kartik, 2007; Yoon and Alexander, 2008; Hongzhi et al., 2012).

- **Accuracy:** A good RS should always check the accuracy level of the items recommended. To what extent the item is accurate will make the RS system run more smoothly.
- **Similarity:** This area emphasizes the fact, how much the product is similar to the users' past interest. There are various algorithms used in the RS system to find the similarity between users or items.
- **Diversity:** The same kind of products should never be recommended to the user on a regular basis, as this lesser down the interest of the user. A dynamic RS gives more diversity.

- **Long Tail (LT):** Some good items do not come into the top-n recommended list of items due to the smaller number of ratings. This problem provides the recommendations of more popular items only.

8.10.7 ATTACKS

The open nature of CF-based RS makes them prone to attacks known as Shilling attacks. Every RS identifies an item set favored by a certain user termed as the recommendation list for that user. Unscrupulous people use unethical ways to push their product into the Top-n recommendation list or pull down their competitors' product from that list. Hence, every attack is either a push or a nuke attack. To accomplish this, attackers inject fake profiles into the RS and give biased ratings to the items leading to erroneous recommendations. An attacker creates a fake profile in such a way as to remain effective and undetected at the same time. The rating is represented as the m-dimensional vector, where m represents a total number of items in the system. Every attack profile has four subparts (Mobasher et al., 2007):

- **Target Item (I_T):** A singleton item, which is to be pushed or nuked.
- **Selected Items (I_S):** A set of items whose rating is determined by a function based on the type of attack.
- **Filler Items (I_F):** A set of items chosen and rated randomly to copy the behavior of an authentic user.
- **Unrated Items (I_N):** A set of items not rated by the attacker.

The attackers closely follow certain attack models while designing an attack (Mobasher et al., 2005; Mobasher et al.,2015; Kaur and Goel, 2016). The target item is generally rated with the highest or the lowest rating. But the rating functions of the filler and selected items lead to different attack models. Let $\bar{r}$ denote the average rating of the RS over all items and users and let $\bar{r}_i$ denote the average rating of a certain item i. Similarly, let σ be the standard deviation of all ratings value over all users and items, and σ_i the standard deviation value of ratings of an item i. Let, $N(r, \sigma^2)$ denote the Gaussian distribution having mean r and variance σ^2 and $\rho(i)$ be the function based on which the filler items I_F are rated.

8.10.7.1   RANDOM ATTACK

Except for target item, attack profiles are generated based on randomly selected users' ratings from the database which contains information about the distribution of ratings. This attack was first mentioned by Lam and Riedl (2004). The set I_S contains no items, whereas I_F contains randomly picked up items whose ratings are given by the function $N(r, \sigma^2)$ centerd on the overall average rating in the database. In random attack:

$$I_S = \phi \text{ and } \rho(i) = N(\bar{r}, \sigma^2)$$

8.10.7.2   AVERAGE ATTACK

In average attack, target item's mean rating is used to generate an attack profile across all the users for filler items. Like the random attack, the I_S remains empty. The filler items are rated by the function $N(\bar{r}, \sigma^2)$ centered on the average rating σ_i of each item i in the database (Burke et al., 2005). Average attack proves to be more effective than a random attack. In average attack:

$$I_S = \phi \text{ and } \rho(i) = N(\bar{r}_i, \sigma_i^2)$$

8.10.7.3   BANDWAGON ATTACK

In bandwagon attack, high ratings are added to generate profiles for the selected item to increase the ratings of popular items. Here, $I_S = \{popular\ items\}$ and $\rho(i) = N(\bar{r}, \sigma^2)$. The items in the I_S are assigned high ratings. This attack needs an additional knowledge about the most popular items in an RS.

8.10.7.4   SEGMENT ATTACK

It is created to increase the recommendations of the set of target items for a certain set of users. Here, I_S is the items similar to target items and $\rho(i) = N(\bar{r}, \sigma_i^2)$. The items in I_S are termed as segment items that are well-liked by the target users. Like the target items, the segment items in I_S are given high ratings while the filler items are given low ratings.

8.11   A SURVEY ON COLLABORATIVE FILTERING (CF)

CF is at the heart of the RSs. It has been employed to develop recommendation techniques that suggest the best suitable items for customers. Yang et al. (2014) have presented a survey on the CF-based RSs categorizing them into social recommendation approaches using matrix factorization and neighborhood-based methods. They have also proposed the idea of utilizing the information from social networks as an additional input in CF for better quality recommendations. Elahi et al. (2016) have discussed the two most popular rating prediction algorithms used in a CF-neighborhood-based model and latent factor model while throwing some light on the cold start problem faced by CF techniques. Instead of using the entire data for CF and also introduces the use of active learning which involves obtaining high-quality data that can better represent a user's preferences, as a solution to the problem of a cold start. An excellent comparison of the CF algorithms found in the literature has been made by Cacheda et al. (2011) using several evaluation metrics, presenting the merits and demerits of every technique. To deal with sparse datasets, a new CF algorithm has been introduced in (Cacheda et al., 2011) that focus on the differences between the items or users rather than looking at their similarities. Shi et al. (2014) have presented a brief review of CF explaining the traditional memory-based and model-based CF approaches in detail. In addition, it also surveys some extended CF algorithms that make use of different information sources apart from the user-item matrix and presents the challenges faced by them. From the perspective of e-vendors in the domain of e-commerce, Karimova et al. (2007) have presented a literature review of the various recommendation techniques including the CF approaches. The analysis reveals the limitations of CF such as computational complexity, accuracy, and so on. An excellent survey of CF has been done by Su et al. (Su and Khoshgoftaar, 2009) where the three categories memory-based, model-based and hybrid CF algorithms have been studied in detail. The strengths and weaknesses of these algorithms, as well as their predictive performances, have been analyzed using several evaluation metrics. Finally, the various challenges faced by CF—scalability, sparsity, synonymy, grey sheep, shilling attacks, and privacy protection, etc. have been presented along with their possible solutions. Nagarnaik and Thomas (2015) have surveyed the various recommendation techniques including the CF algorithms explaining its various categories. A literature review has been done on the techniques that have been proposed to overcome the challenges faced by CF algorithms. Also, a hybrid CF technique has been proposed taking a combination of CF techniques and pattern finding algorithms for a better-quality web page recommendation. Yang et al. (2016) have shown the entire framework of a typical CF-based RS. A detailed survey of the working of CF algorithms-similarity metrics, prediction algorithms, and neighbor selection has been done, and case studies have presented to measure the accuracy of various CF algorithms using evaluation metrics.

8.12 CURRENT RESEARCH TRENDS IN COLLABORATIVE FILTERING (CF) (2011–2017)

Journal articles and conference proceedings were collected from four major electronic databases, i.e., ACM Digital Library, IEEE Explore, Springer, and Science Direct (Elsevier) to find the research trends in CF for the recommendation. The following queries were executed in Google Scholar:

a. (CF in RSs OR (issue OR issues OR challenge OR challenges OR problem OR problems)).
b. (Neighborhood CF in recommendation systems OR (issue OR issues OR challenge OR challenges OR problem OR problems)).
c. (CF OR neighborhood CF OR RSs)

Additional specifications about papers were also added in the advanced search of Google Scholar, to get more filtered papers out:

a. 2011–2017 is selected in the field of date section.
b. IEEE OR Elsevier OR ACM OR Springer has-been selected in the "published in" field.

The application of the above queries yielded around 500 research papers. Of the 500 research papers, we selected only 277 papers which are related to the area of computers and its allied fields. The keywords and abstract of each paper were used in the categorization of the collected papers, which led to the following results as shown in Figures 8.4–8.6. Figure 8.4 depicts the papers as well as their distribution among the top publications on selected 277 papers of CF. Out of the 277 papers, 77 are model-based, 162 are neighborhood-based, and the rest apply some other filtering technique in addition to CF as shown in Figure 8.5. And finally, Figure 8.6 states that the 162 papers containing neighborhood-based CF can be further segregated in a number of sub-domains based on the problems they try to rectify.

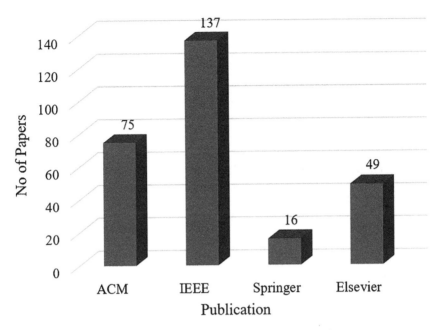

FIGURE 8.4 Number of paper distribution of CF, in each publication.

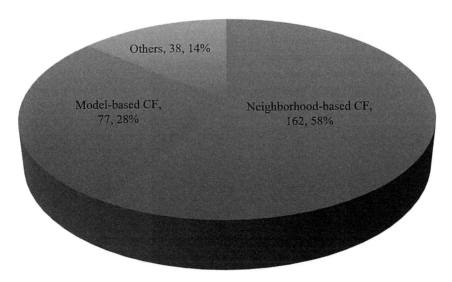

FIGURE 8.5 Number of papers in different categories of CF.

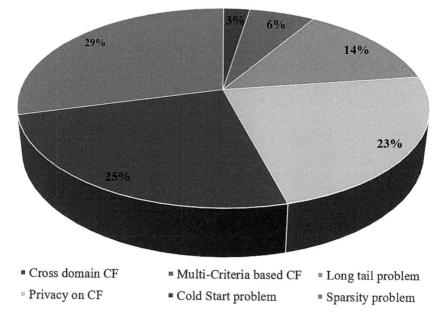

■ Cross domain CF   ■ Multi-Criteria based CF   ■ Long tail problem
■ Privacy on CF   ■ Cold Start problem   ■ Sparsity problem

FIGURE 8.6   The percentage share of published papers in different categories of neighborhood-based CF.

8.13   THE FUTURE OF COLLABORATIVE FILTERING (CF) BASED RECOMMENDER SYSTEMS (RSS)

The accuracy of neighborhood-based CF mainly depends upon the top-n list of similar users or items (Yi et al., 2019; Soojung, 2019). Recommendations using these similar users/items tend towards more popular items. But an ideal RS has different properties such as more personalization, more diverse, more serendipity, and more novel. Recommendations using neighborhood-based CF can be improved if the researchers use information from social networks and contextual information of user or item with their rating information (Ambulgekar et al., 2019).

8.14   CONCLUSION

RS has found its usefulness in several fields of e-services. Among several filtering approaches used in RSs, CF is most common and popular. The mainly used perception of CF-based is that the rating of the items given by similar users will be close and similar items have similar rating patterns. On this basis, CF suggests the recommendable items to the users. CF extracts user ratings either implicitly or explicitly. To find the user-user and item-item similarity, several metrics are used to find the similarity of user-user and item-item. The similarity value has been used to predict the recommendable items. Several prediction approaches such as MC, WA, Z score (ZS), etc. are used. Different performance metrics such as MAE, RMSE, Coverage, etc. measure the correctness of recommendation. For effective implementation of CF-based RS, several challenges need to be addressed. New user and a new item, sparsity, scalability, synonymy, and LT problems are among them. Security and trust attacks on RS are also a major concern in people's acceptance of RSs. Due to the utility of CF in RS; it has attracted the attention of the academicians and researchers to make it more effective. The future of CF-based RS will be more personalized and diverse with more serendipity.

KEYWORDS

- **data extraction**
- **e-commerce**
- **future direction**
- **neighborhood-based collaborative filtering**
- **performance metrics**
- **recommendation system**
- **recommender system attacks**
- **research trends**
- **similarity metrics**

REFERENCES

ACRnews, (2017). [Online]. Available: http://www.acr-news.com/ (accessed on 16 February 2020).

Ali, K., & Van Stam, W., (2004). "TiVo: Making show recommendations using a distributed collaborative filtering architecture." In: *Proceedings of the Tenth ACM SIGKDD International Conference on Knowledge Discovery and Data Mining.*

Alloway, T., (2018). *"Amazon Urges California Referendum on Online Tax."* [Online]. Available: https://ftalphaville.ft.com/2011/07/12/619451/amazon-urges-california-referendum-on-online-tax/ (accessed on 16 February 2020).

Ambulgekar, H. P., Manjiri, K. P., & Kokare, M. B., (2019). "A survey on collaborative filtering: Tasks, approaches and applications." In: *Proceedings of International Ethical Hacking Conference* (pp. 289–300).

Amolochitis, E., Christou, I. T., & Tan, Z. H., (2014). "Implementing a commercial-strength parallel hybrid movie recommendation engine." *IEEE Intelligent Systems, 29,* 92–96.

Asnicar, F., & Tasso, C., (1997). "ifWeb: A prototype of user model-based intelligent agent for document filtering and navigation in the world wide web." In: *Proceedings of 6th International Conference on User Modeling.*

Avesani, P., Massa, P., & Tiella, R., (2005). "Moleskiing.it: A Trust-aware Recommender System for Ski Mountaineering." In: *Proceedings of the ACM Symposium on Applied Computing.*

Balabanović, M., & Shoham, Y., (1997). "Fab: Content-based, collaborative recommendation." *Communications of the ACM, 40,* pp. 66–72.

Baltrunas, L., Ludwig, B., Peer, S., & Ricci, F., (2012). "Context relevance assessment and exploitation in mobile recommender systems." *Personal and Ubiquitous Computing, 16*(5), 507–526.

Baraglia, R., Frattari, C., Muntean, C. I., Nardini, F. M., & Silvestri, F., (2012). "RecTour: A recommender system for tourists." In: *Proceedings of the 2012 IEEE/WIC/ACM International Joint Conference on Web Intelligence and Intelligent Agent Technology.*

BBC, (2018). *"Orlando Figes to Pay Fake Amazon Review Damages."* [Online]. Available: https://www.bbc.com/news/uk-10670407 (accessed on 16 February 2020).

Biletskiy, Y., Baghi, H., Keleberda, I., & Fleming, M., (2009). "An adjustable personalization of search and delivery of learning objects to learners." *Expert Systems with Applications, 36*(5), pp. 9113–9121.

Bilge, A., & Kaleli, C., (2014). "A multi-criteria item-based collaborative filtering framework." In: *11th International Joint Conference on Computer Science and Software Engineering.*

Billsus, D., & Pazzani, M. J., (2000). "User modeling for adaptive news access." *User Modeling and User-Adapted Interaction, 10,* pp. 147–180.

Bjelica, M., (2010). "Towards TV recommender system: Experiments with user modeling." *IEEE Transactions on Consumer Electronics, 56*(3), 1763–1769.

Blanco-Fernández, Y., Arias, J. J. P., Nores, M. L., Gil-Solla, A., & Cabrer, M. R., (2006). "AVATAR: An improved solution for personalized TV based on semantic inference." *IEEE Transactions on Consumer Electronics, 52*(1), pp. 223–231.

Bobadilla, J., Hernando, A., Ortega, F., & Abraham, G., (2012). Collaborative filtering based on significances. *Information Sciences, 185*(1), pp. 1–17.

Boutet, A., Frey, D., Guerraoui, R., Jegou, A., & Kermarrec, A. M., (2013). "WhatsUp: A decentralized instant news recommender." In: *IEEE 27th International Symposium on Parallel Distributed Processing.*

Braunhofer, M., Kaminskas, M., & Ricci, F., (2013). "Location-aware music recommendation." *International Journal of Multimedia Information Retrieval, 2*(1), 31–44.

Burke, R. D., Hammond, K. J., & Young, B. C., (1996). "Knowledge-based navigation of complex information spaces." In: *Proceedings of the Thirteenth National Conference on Artificial Intelligence and Eighth Innovative Applications of Artificial Intelligence Conference (AAAI)*. Portland, Oregon.

Burke, R., (1999). "The wasabi personal shopper: A case-based recommender system." In: *Proceedings of the 11ᵗʰ National Conference on Innovative Applications of Artificial Intelligence.*

Burke, R., (2002). "Hybrid recommender systems: Survey and experiments." *User Modeling and User-Adapted Interaction, 12*(4), pp. 331–370.

Burke, R., Mobasher, B., Bhaumik, R., & Williams, C., (2005). "Segment-based injection attacks against collaborative filtering recommender systems." In: *Fifth IEEE International Conference on Data Mining (ICDM'05).*

Cacheda, F., Carneiro, V., Fern'andez, D., & Formoso, V., (2011). "Comparison of collaborative filtering algorithms: Limitations of current techniques and proposals for scalable, high-performance recommender systems." In: *TWEB* (Vol. 5, No. 1/2, p. 33).

Cao, Y., & Li, Y., (2007). "An intelligent fuzzy-based recommendation system for consumer electronic products." *Expert Systems with Applications, 33*(1), pp. 230–240.

Capuano, N., Gaeta, M., Ritrovato, P., & Salerno, S., (2014). "Elicitation of latent learning needs through learning goals recommendation." *Computers in Human Behavior, 30*, pp. 663–673.

Celma, Ò., & Serra, X., (2008). "FOAFing the music: Bridging the semantic gap in music recommendation." *Web Semantics: Science, Services and Agents on the World Wide Web, 6*(4).

Chen, C. M., & Duh, L. J., (2008). "Personalized web-based tutoring system based on fuzzy item response theory." *Expert Systems with Applications, 34*, pp. 2298–2315.

Chen, C. M., Duh, L. J., & Liu, C. Y., (2004). "A personalized courseware recommendation system based on fuzzy item response theory." In: *IEEE International Conference on e-Technology, e-Commerce and e-Service.*

Chen, L., & Pu, P., (2012). "Critiquing-based recommenders: Survey and emerging trends." *User Modeling and User-Adapted Interaction, 22*(1), pp. 125–150.

Chesnevar, C. I., & Maguitman, A. G., (2004). "ArgueNet: An argument-based recommender system for solving Web search queries." In: *2ⁿᵈ International IEEE Conference on Intelligent Systems.*

Christou, I. T., Amolochitis, E., & Tan, Z. H., (2016). "AMORE: design and implementation of a commercial-strength parallel hybrid movie recommendation engine." *Knowledge and Information Systems, 47*(3), 671–696.

Cobos, C., Rodriguez, O., Rivera, J., Betancourt, J., Mendoza, M., Leó, N. E., & Herrera-Viedma, E., (2013). "A hybrid system of pedagogical pattern recommendations based on singular value decomposition and variable data attributes." *Information Processing and Management: An International Journal, 49*, 607–625.

Console, L., Torre, I., Lombardi, I., Gioria, S., & Surano, V., (2003). "Personalized and adaptive services on board a car: An application for tourist information." *Journal of Intelligent Information Systems, 21*(3), pp. 249–284.

Cornelis, C., Lu, J., Guo, X., & Zhang, G., (2007). "One-and-only item recommendation with fuzzy logic techniques." *Information Sciences, 177*, pp. 4906–4921.

Daniel, M. F., & Kartik, H., (2007). "Recommender systems and their impact on sales diversity." In: *Proceedings of the 8ᵗʰ ACM Conference on Electronic Commerce (EC '07)* (pp. 192–199). ACM, New York, NY, USA.

Deshpande, M., & Karypis, G., (2004). "Item-based top-n recommendation Algorithms." *ACM Transactions on Information Systems (TOIS), 22*, pp. 143–177.

Domingues, M., Gouyon, F., Jorge, A., Leal, J., Vinagre, J., Lemos, L., & Sordo, M., (2013). "Combining usage and content in an online recommendation system for music in the long tail." *International Journal of Multimedia Information Retrieval, 2*(1), 3–13.

Elahi, M., Ricci, F., & Rubens, N., (2016). "A survey of active learning in collaborative filtering recommender systems." In: *Computer Science Review* (Vol. 20, pp. 29–50).

Esteban, B., Tejeda-Lorente, Á., Porcel, C., Arroyo, M., & Herrera-Viedma, E., (2014). "TPLUFIB-WEB: A fuzzy linguistic Web system to help in the treatment of low back pain problems." *Knowledge-Based Systems, 67*, pp. 429–438.

Farzan, R., & Brusilovsky, P., (2006). "Social navigation support in a course recommendation system." In: *Adaptive Hypermedia and Adaptive Web-Based Systems: 4th International Conference (AH 2006)*. Dublin, Ireland.

Fesenmaier, D. R., Ricci, F., Schaumlechner, E., Wöber, K., & Zanella, C., (2003). "DIETORECS: Travel advisory for multiple decision styles." In: *Proceedings of the International Conference on Information and Communication Technologies in Tourism*. Wien, Austria.

FoxTrit, (2017). [Online]. Available: http://www.foxtrot.com/wp-content/endurance-page-cache/_index.html (accessed on 16 February 2020).

García-Crespo, A., Chamizo, J., Rivera, I., Mencke, M., Colomo-Palacios, R., & Gómez-Berbís, J. M., (2009). "SPETA: Social pervasive e-tourism advisor." *Telematics and Informatics, 26*, pp. 306–315.

García-Crespo, Á., López-Cuadrado, J. L., Colomo-Palacios, R., González-Carrasco, I., & Ruiz-Mezcua, B., (2011). "Sem-Fit: A semantic based expert system to provide recommendations in the tourism domain." *Expert Systems with Applications, 38*, pp. 13310–13319.

Garfinkel, R., Gopal, R., Tripathi, A., & Yin, F., (2006). "Design of a shopbot and recommender system for bundle purchases." *Decision Support Systems, 42*(3), pp. 1974–1986.

Guo, X., and Lu, J., (2007). "Intelligent e-government services with personalized recommendation techniques." *International Journal of Intelligent Systems, 22*, pp. 401–417.

Hauver, D., & French, J., (2001). "Flycasting: using collaborative filtering to generate a playlist for online radio." In: *First International Conference on Web Delivering of Music*.

Hayes, C., & Cunningham, P., (2001). "Smart radio-community based music radio." *Knowledge Based Systems, 14*, pp. 197–201.

Herlocker, J. L., Konstan, J. A., Borchers, A., & Riedl, J., (1999). "An algorithmic framework for performing collaborative filtering." In: *22nd Annual International ACM SIGIR Conference on Research and Development in Information Retrieval*.

Hongzhi, Y., Bin, C., Jing, L., Junjie, Y., & Chen, C., (2012). "Challenging the long tail recommendation." In: *Proc. VLDB Endow., 5*(9), 896–907.

Hotho, A., Jäschke, R., Schmitz, C., & Stumme, G., (2006). "Information retrieval in folksonomies: Search and ranking." In: *The Semantic Web: Research and Applications: 3rd European Semantic Web Conference*. Budva, Montenegro.

Hu, J., Wang, B., Liu, Y., & Li, D. Y., (2012). "Personalized tag recommendation using social influence." *Journal of Computer Science and Technology, 27*(3), pp. 527–540.

IMDb, (2017). [Online]. Available: http://www.imdb.com/ (accessed on 16 February 2020).

Jäschke, R., Marinho, L. B., Hotho, A., Schmidt-Thieme, L., & Stumme, G., (2007). "Tag Recommendations in Folksonomies." In: *11th European Conference on Principles and Practice of Knowledge Discovery in Databases*. Warsaw, Poland.

Jinni, (2017). [Online]. Available: http://www.jinni.com/ (accessed on 16 February 2020).

Karimova, F. (2016), "A Survey of e-Commerce Recommender Systems," *European Scientific Journal, 12*(34), 75–89.

Kaur, P., & Goel, S., (2016). "Shilling attack models in recommender system." In: *International Conference on Inventive Computation Technologies (ICICT)* (Vol. 2, pp. 1–5).

Konstan, J. A., Miller, B. N., Maltz, D., Herlocker, J. L., Gordon, L. R., & Riedl, J., (1997). "GroupLens: Applying collaborative filtering to Usenet news." *Communications of the ACM, 40*, pp. 77–87.

Lakshmi, S. S., & Lakshmi, T. A., (2014). "Recommendation systems: Issues and challenges." In: *International Journal of Computer Science and Information Technologies, 5*.

Lam, S. K., & Riedl, J., (2004). "Shilling recommender systems for fun and profit." In: *Proceedings of the 13ᵗʰ International Conference on World Wide Web* (pp. 393–402).

Lawrence, R., Almasi, G., Kotlyar, V., Viveros, M., and Duri, S., (2001). "Personalization of supermarket product recommendations." *Data Mining and Knowledge Discovery, 5*(1), pp. 11–32.

Lee, S. K., Cho, Y. H., & Kim, S. H., (2010). "Collaborative filtering with ordinal scale-based implicit ratings for mobile music recommendations." *Information Sciences, 180*(11), pp. 2142–2155.

Lei, S., (2013). "Trading-off among accuracy, similarity, diversity, and long-tail: A graph-based recommendation approach." In: *Proceedings of the 7ᵗʰ ACM Conference on Recommender Systems*.

Levandoski, J. J., Sarwat, M., Eldawy, A., & Mokbel, M. F., (2012). "LARS: A location-aware recommender system." In: *Proceedings of the 2012 IEEE 28ᵗʰ International Conference on Data Engineering*.

Li, D., Miao, C., Chu, S., Mallen, J., Yoshioka, T., & Srivastava, P., (2018). *"Stable Matrix Approximation for Top-n Recommendation on Implicit Feedback Data."* In Hawaii International Conference on System Sciences 2018 (HICSS-51).

Liu, Q., & Karger, D. R., (2015). "Kibitz: End-to-end recommendation system builder." In: *RecSys*.

Lu, J., (2004). "Personalized e-learning material recommender system." In: *Proceedings of International Conference on Information Technology for Application*.

Lu, J., Shambour, Q., Xu, Y., Lin, Q., & Zhang, G., (2010). "BizSeeker: A hybrid semantic recommendation system for personalized government-to-business e-services." *Internet Research, 20*, pp. 342–365.

Luz, N., Moreno, M., Anacleto, R., Almeida, A., & Martins, C., (2013). "A hybrid recommendation approach for a tourism system." *Expert Systems with Applications, 9*(40), 3532–3550.

Marcel, M. A., Ball, M., Boley, H., Greene, S., Howse, N., Lemire, D., & Mcgrath, S., (2003). "RACOFI: A rule-applying collaborative filtering system." In: *Proc. IEEE/WIC COLA'03*. Halifax, Canada.

Martín-Vicente, M. I., Gil-Solla, A., Ramos-Cabrer, M., Blanco-Fernández, Y., & Servia-Rodríguez, S., (2012). "Semantics-driven recommendation of coupons through digital TV: Exploiting synergies with social networks." In: *IEEE International Conference on Consumer Electronics*.

Martínez, L., Rodríguez, R. M., & Espinilla, M., (2009). "REJA: A georeferenced hybrid recommender system for restaurants." In: *Proceedings of the 2009 IEEE/WIC/ACM*

International Joint Conference on Web Intelligence and Intelligent Agent Technology (pp. 187–190).

Mccarthy, K., Reilly, J., Mcginty, L., & Smyth, B., (2004). "Thinking positively-explanatory feedback for conversational recommender systems." In: *Proceedings of the ECCBR 2004 Workshops.*

Melville, P., Mooney, R. J., & Nagarajan, R., (2002). "Content-boosted collaborative filtering for improved recommendations." In: *Eighteenth National Conference on Artificial Intelligence.* Edmonton, Alberta, Canada.

Meo, P. D., Quattrone, G., & Ursino, D., (2008). "A decision support system for designing new services tailored to citizen profiles in a complex and distributed e-government scenario." *Data and Knowledge Engineering, 67,* pp. 161–184.

Miller, B. N., Konstan, J. A., & Riedl, J., (2004). "PocketLens: Toward a personal recommender system." *ACM Transactions on Information Systems, 22*(3), pp. 437–476.

Mobasher, B., Burke, R., Bhaumik, R., & Sandvig, J. J., (2007). "Attacks and remedies in collaborative recommendation." *IEEE Intelligent Systems, 22*(3), 56–63.

Mobasher, B., Burke, R., Bhaumik, R., & Williams, C. (2007). "Toward trustworthy recommender systems: An analysis of attack models and algorithm robustness." *ACM Trans. Internet Technol., 7*(4).

Mobasher, B., Burke, R., Bhaumik, R., & Williams, C., (2005). "Effective attack models for shilling item-based collaborative filtering systems." In: *Proceedings of the WebKDD Workshop, Held in Conjunction with ACM SIGKDD2005.*

Moreno, A., Valls, A., Isern, D., Marin, L., & Borràs, J., (2013). "SigTur/E-destination: Ontology-based personalized recommendation of tourism and leisure activities." *Engineering Applications of Artificial Intelligence, 26*(1), pp. 633–651.

Moukas, A., & Maes, P., (1998). "Amalthaea: An evolving multi-agent information filtering and discovery system for the WWW." *Autonomous Agents and Multi-Agent Systems, 1*(1), 59–88.

Movielens, (2017). [Online]. Available: https://movielens.org/ (accessed on 16 February 2020).

Nagarnaik, P., & Thomas, A., (2015). "Survey on recommendation system methods." In: *2nd International Conference on Electronics and Communication Systems (ICECS)* (pp. 1603–1608).

nanoCROWD, (2017). [Online]. Available: http://nanocrowd.com/ (accessed on 16 February 2020).

Nanopoulos, A., Rafailidis, D., Symeonidis, P., & Manolopoulos, Y., (2010). "Music box: Personalized music recommendation based on cubic analysis of social tags." *IEEE Transactions on Audio, Speech and Language Processing, 18*(2), pp. 407–412.

Natarajan, N., Shin, D., & Dhillon, I. S., (2013). "Which app will you use next?: Collaborative filtering with interactional context." In: *Proceedings of the 7th ACM Conference on Recommender Systems.*

Nguyen, T. T. S., Lu, H. Y., & Lu, J., (2014). "Web-page recommendation based on web usage and domain knowledge." *IEEE Transactions on Knowledge and Data Engineering, 26*(10), pp. 2574–2587.

Oh, J., Kim, S., Kim, J., & Yu, H., (2014). "When to recommend: A new issue on TV show recommendation." *Information Sciences, 280,* pp. 261–274.

Oscar, C., (2010). "*Music Recommendation and Discovery: The Long Tail, Long Fail, and Long Play in the Digital Music Space.*" Springer Publishing Company, Incorporated.

Pampın, H. J. C., Jerbi, H., & O'Mahony, M. P. (2015), "Evaluating the relative performance of collaborative filtering recommender systems, *Journal of Universal Computer Science*, *21*(13), 1849–1868.

Park, Y. J., (2013). "An adaptive match-making system reflecting the explicit and implicit preferences of users." *Expert Systems with Applications: An International Journal*, *40*, 1196–1204.

Park, Y. J., & Tuzhilin, A. (2008, October), "The long tail of recommender systems and how to leverage it," In *Proceedings of the 2008 ACM Conference on Recommender Systems* (pp. 11–18).

Parra, D., Brusilovsky, P., & Trattner, C., (2014). "See what you want to see: Visual user-driven approach for hybrid recommendation." In: *Proceedings of the 19ᵗʰ International Conference on Intelligent User Interfaces*.

Pashtan, A., Blattler, R., Andi, A. H., & Scheuermann, P., (2003). "CATIS: A context-aware tourist information system." In: *Proceedings of the 4ᵗʰ International Workshop of Mobile Computing*.

Poonam, T. B., Goudar, R. M., & Sunita, B., (2015). "Article: Survey on collaborative filtering, content-based filtering and hybrid recommendation system." *International Journal of Computer Applications*, 31–36.

Porcel, C., & Herrera-Viedma, E., (2010). "Dealing with incomplete information in a fuzzy linguistic recommender system to disseminate information in university digital libraries." *Knowledge-Based Systems, 23*, pp. 32–39.

Porcel, C., Herrera-Viedma, E., & Moreno, J. M., (2009). "A multi-discipliner recommender system to advice research resources in university digital libraries." *Expert Systems with Applications*, *36*, pp. 12520–12528.

Porcel, C., López-Herrera, A. G., & Herrera-Viedma, E., (2009). "A recommender system for research resources based on fuzzy linguistic modeling." *Expert Systems with Applications: An International Journal*, *36*, pp. 5173–5183.

Pratikshashiv, (2015). "*Flipkart Uses Collaborative Based Filtering.*" [Online]. Available: https://pratikshashiv.wordpress.com/ (accessed on 16 February 2020).

Renda, M. E., & Straccia, U., (2005). "A personalized collaborative digital library environment: A model and an application." *Information Processing and Management: An International Journal*, *41*, 5–21.

Rikitianskii, A., Harvey, M., & Crestani, F., (2014). "A personalized recommendation system for context-aware suggestions." In: *Advances in Information Retrieval: 36ᵗʰ European Conference on IR Research*. ECIR.

Rotten Tomatoes, (2017). [Online]. Available: https://www.rottentomatoes.com/ (accessed on 16 February 2020).

Ruotsalo, T., Haav, K., Stoyanov, A., Roche, S., Fani, E., Deliai, R., Mäkelä, E., Kauppinen, T., & Hyvönen, E., (2013). "Smart museum: A mobile recommender system for the web of data." *Web Semantics: Science, Services and Agents on the World Wide Web*, *20*.

Samundeeswary, K., & Krishnamurthy, V. (2017, June), "Comparative study of recommender systems built using various methods of collaborative filtering algorithm." In *2017 International Conference on Computational Intelligence in Data Science (ICCIDS)* (pp. 1–6). IEEE.

Salter, J., & Antonopoulos, N., (2006). "Cinema screen recommender agent: Combining collaborative and content-based filtering." *IEEE Intelligent Systems*, *21*(1), pp. 35–41.

CHAPTER 9

Business Application Analytics Through the Internet of Things

DEEPAK KUMAR SHARMA, PRAGI MALHOTRA, and JAYANT VERMA

Division of Information Technology, Netaji Subhas University of Technology, (Formerly Netaji Subhas Institute of Technology), New Delhi, India, E-mails: dk.sharma1982@yahoo.com (D. K. Sharma), pragi18@gmail.com (P. Malhotra), jayantverma1998@gmail.com (J. Verma)

ABSTRACT

Internet of things (IoT) is an upcoming technology that helps connect various types of devices, mechanical, and digital machines with the help of appropriate hardware and sensors, which as a whole, are able to transfer data without the requirement of human intervention. Business applications are softwares designed for the purpose of performing or improving business processes (BP). More specifically analytic applications make use of historical data about the different operations in business and equip business users with information and instruments that allow them to make the business functions better and more efficient.

Through IoT, the process of business analytics (BA) is made easier and diversified since real-time analytics is possible with the use of IoT. There exist promising technologies that allow smooth and unified access to many IoT resources that are diverse and distributed in nature. Technology behind resource-oriented computing is explored in this chapter. This approach provides an architecture where the front end is incorporated into the back end IoT devices which deal with applications that deal with BP. This architecture has many desirable features like a dynamic and responsive service replacement facility, ease of access to IoT services for programmers and execution of IoT aware BP in a distributed manner.

This chapter recognizes the need to devise a business model framework for IoT services in order to gain profits and discusses various business models and chooses a framework for the same for IoT. We further discuss how IoT offers gains in the various domain of business. To sum all this up, we present case studies to provide readers with a better understanding of all the areas covered. We also mention the challenges faced in the current scenario.

9.1 INTRODUCTION

9.1.1 WHAT IS IoT

Internet of things (IoT) refers to the ever-growing network of real-world physical objects that feature an IP address for internet connectivity, and the communication that occurs between these objects and other Internet-enabled devices and systems. In other words, any stand-alone device which can be connected to the internet which can be monitored and/or operated from a remote location is considered an IoT device. IoT ecosystem comprises of components like remotes, dashboards, networks, analytics, data storage, gateways, and security that enable businesses, governments, and consumers to connect to their IoT devices.

9.1.2 SCOPE OF IoT

Scope of IoT is vast; it is being embedded in almost everything these days. The main reason to shift to IoT is efficiency and better control over different steps of the operation. Industries affected by IoT are as follows and their market shares are depicted in Figure 9.1:

1. Manufacturing;
2. Transportation;
3. Defense;
4. Agriculture;
5. Retail;
6. Infrastructure;
7. Logistics;
8. Banks;
9. Connected homes and smart buildings;

10. Healthcare;
11. Smart cities.

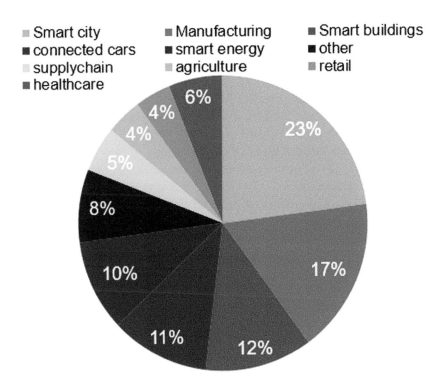

FIGURE 9.1   Market share of different industries in IoT.

9.1.3   *ARCHITECTURE OF IoT*

1. **Entity:** This consists of governments, businesses, and consumers who are using IoT.
2. **Physical Layer:** Comprises of the hardware components that are required in an IoT device, including the RFID chip, sensors, networking equipment, etc.
3. **Device Layer:** Its main responsibility is to transmit the data that is collected by the physical layer to different devices. It uses different protocols and routing methods to achieve successful networking.

4. **Application Layer:** Comprises of the protocols and interfaces that are used by devices to identify and communicate with each other.
5. **Dashboard:** Displays useful information about the IoT ecosystem to the end-users and a platform through which they can monitor/control their IoT ecosystem.
6. **Analytics:** The data generated by the IoT devices needs to be analyzed to obtain usable information from it, Analytics is the department which handles this job. Softwares are used to collect and process the data. This is useful in generating reports, making predictions, and keeping an overall check on the entire process.
7. **Data Storage:** The data generated by the IoT devices needs to be stored, data processing can't be done in real time due to the complexity of softwares hence it becomes essential to include a data storage unit.
8. **Network Layer:** It is the network layer through which the entities communicate with the devices; data transfer among the devices is possible through the network layer. Data transfer here refers to the transfer that takes place both ways.

9.2 BUSINESS PROCESSES (BP)

9.2.1 CHANGING DYNAMICS OF BUSINESS

What is business? The answer to this question has been gradually changing and this change can be seen in a series of defined eras which last about 50 years. These eras can be characterized by a set of unique features and some companies that represent that era for instance, Standard oil at the turn of the last century.

The dominant standard under which businesses have been operating is starting to shift. The main cause of this shift can be attributed to reasons like rapid advancements in technology, a wider and more diverse consumer base, power shift from top-level managers to the people who actually deliver value to customers, an increase in the number of competitors. The traditional way of doing business, where the businessmen are simply concerned with making goods or services and maintaining profits is no longer sufficient to survive in this new era of business.

9.2.2 TYPES/MODEL

Businesses can be of many types and among these there are four major ones. They are:

1. **Service Business:** A service type of business provides products that are not concrete. These types of firms offer professional expertise, knowledge, guidance, and similar type of products. Some examples of these businesses are: IT firms, NBFCs, parlors, Auto repair shops.

2. **Merchandising Business:** A merchandising type of business buys and sells products at different prices. The selling price is higher than the purchase cost and this is how these businesses make profit. Examples are: convenience stores and other resellers.

3. **Manufacturing Business:** In this type of business, also products are bought at a cost but the difference between this and merchandising business is that here the products are used to produce some new product whereas earlier the product was simply sold.

4. **Hybrid Business:** These are companies that can be categorized in more than one type of business. A cafe, for example, sells a bottle of champagne (merchandising), combines food ingredients in making a meal (manufacturing), and takes customer orders (service).

9.2.3 BUSINESS ANALYTICS (BA)

It is the method of repetitive, systematic assessment of a business's data, in which statistical analysis has special importance. This tool facilitates the process of making decisions that are based on data. BP can be automated and optimized with the help of BA. Companies that focus on data and give it a lot of importance and they treat like as an asset and use it to gain competitive edge. The success of BA is dependent upon the quality of data; expertise of experts who are well versed with the technical know-how as well the business. BA techniques can be divided into two broad categories. Basic business intelligence is the first area. The main aim here is to analyze historical data and to anticipate the performance of a business department or a member of the staff or a team over a specific period of time. This practice is quite common, mature, and most of the enterprises today have tools and means required to use this analysis. The second area includes statistical analysis on a deeper level. Statistical algorithms can be applied to historical data in order to perform analysis and then forecast about future performance about a product or predictions about the implications caused by changing website design.

9.2.4   NEED FOR IoT IN BUSINESS APPLICATIONS

As mentioned above, the fundamentals of business are changing and in order to keep up with the competitors, it is necessary for every business to embrace the new technologies. IoT is a relatively new field but possible advantages of incorporating it into the business application analytics are plenty.

The people who manufacture IoT devices have a huge advantage since they can directly access their customers' behavior. Now, this data helps them to study various customer characteristics which can further be exploited to expand business. Customer characteristics, for example, can include their shopping preferences and these can be used to decide which products need to be discounted and to create schemes that cater to the needs of customers. IoT will help to strengthen the business operations. Routine tasks do not require much expertise to perform but are nonetheless very necessary. To save up on time IoT devices can be used to automate these redundant tasks. Managing inventory can be made more efficient with the help of IoT. For example, if someone owns a grocery store and some of the items in inventory is about to perish then the owners are automatically informed.

9.3   INTEGRATION OF IoT IN BUSINESS APPLICATION

9.3.1   RESOURCE ORIENTED INTEGRATION ARCHITECTURE FOR IoT (REST INSPIRED ARCHITECTURE)

This section explores the technical aspect of integrating IoT into BP. To transform the idea of incorporation of IoT in BA into reality there is a need for a common underlying architecture. A uniform architecture can help create an easy as well as standardized access to IoT devices. We study the resource-oriented method which provides an architecture where the front end is incorporated into the back end IoT devices which deal with applications that deal with BP. The major characteristics of this architecture are:

- Access to IoT services which is very programmer friendly.
- Replacement facility in case of failure.
- Execution of IoT in such a manner that it is also aware of BP in a decentralized manner.
- The necessary APIs to invoke IoT services, thereby facilitating BP developers.
- Event-based model based on publish-subscribe mechanism.

Therefore, we can draw the conclusion that the aim of this IoT architecture is to facilitate integration of IoT resources into the simulated BP world. The resource-oriented approach (ROA) approach can be employed for the construction of a blueprint and execution of a standard architectural model that provides the main constituents needed for start-to-end incorporation of IoT systems with emphasis on BPs that are IoT-aware. This approach employs the principles of representational state transfer (REST) design philosophy. These principles when applied to IoT environment can lead to the formation of environments where sensing, communication, and computing can be easily amalgamated. A small example in the domain of automobile industry is presented first in order to better explain what exactly the requirements for such architecture are. In this example, the system under consideration is built for the purpose of transferring real time information about the vehicles to the manufacturers in order to facilitate timely and easy maintenance and repair. The whole BP is simply a combination of local jobs put together in order to deliver the underlying reasoning behind the application. It is divided into three parts:

1. **Set Top Box (STB):** It locally composes and IoT devices collect data for it. The part of BP present on STB transfers the data collected and the outcome generated to remote BP.
2. **Remote Computer:** This provides prompt analysis and vehicle supervision in the long term. Automobile data is then available to the engineers throughout routine check-ups. Emergency situations like low brake fluid are transmitted to engineers via real-time communication channels.
3. **Smart Phone:** The information needs to transmit to the smart-phones also in order to inform the customers about the analysis and subsequent actions recommended by manufacturers.

Figure 9.2 diagrammatically represents the process mentioned above.

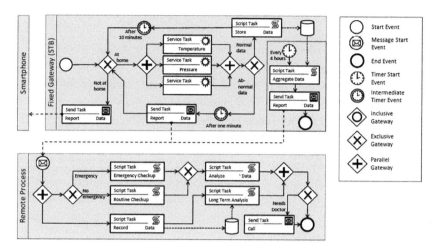

FIGURE 9.2 Performing a BP using STB, remote computer and smart. (Reprinted with permission from Dar, et al., 2014. © 2014 Elsevier.)

This application scenario gives an insight into the design requirements of the needed architecture. These principles are:

- **Unified Integration:** One of the most basic requirements for the services provided by IoT devices is that they should be available through web, thereby rendering these services available everywhere. Thus, the architecture must support devices such as web services (WS) that are accessible through a unique uniform resource identifier (URIs). In addition to this, there is also a need for a uniform service definition language which can describe the non-functional as well as functional aspects and can be used to attain interoperability between platforms.

- **Support for Event-Based Interaction:** It is very evident from the example above that the nature of IoT applications is dependent on various events. For e.g., if the date of a monthly car maintenance visit is due then an alert needs to be sent, if the tires are experiencing more wear and tear than normal then an analysis report citing reasons and diagnosis is prepared and shared with the customer, if there is some serious problem with the engine then emergency alert is sent. Thus, these are all various kinds of events and the corresponding action taken by the application is based on the event. To ensure proficiency the device shouldn't need to conduct a survey to detect the device level events all the time. So, the battery life is saved since the device reports only when an event occurs.

- **Replacement of Services:** The availability of services delivered by IoT devices is more susceptible to changes than traditional WSs. For e.g., even small events like low battery can lead to unavailability of services. To ensure smooth working of IoT services, there is a need for service replacement facility as well as retainment of the current state of BP so that once the service replacement is finished the process can resume execution.
- **Decentralized Business Process:** In decentralized processes, the global goal is achieved by the operation of lower level components on local information. This way the domain expert can design and at the same time describe the communication between various BPs and also helps overcome challenges in adjusting to ever-changing environment of smart devices.

The ROA architecture aims to fulfill all these design requirements. Figure 9.3 depicts the same. The services provided (lowermost in fig) are incorporated with the BPs that are IoT aware (topmost in fig) via programming framework and other parts like STB (in the middle). Following is a detailed explanation of all the constituents:

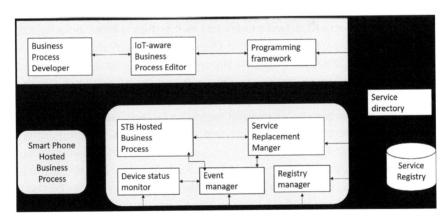

FIGURE 9.3   Constituents of ROA architecture.

1. **Service Description:** Service is functionality that offered set of resources and available to the users via interfaces. Web services description language (WSDL) or web services description language (WSDL) provides the description of services of ROA. Every service has a set of resources and an interface as mentioned

above. The description consists of a name and an URI of the resource where the service endpoint is contained. Resources are described with name, interface type, and URI which gives full access to resource as well as its links to the interfaces. Each interface in turn depicts the methods supported by resource. Service description is stored in the service registry. With the help of this design, many resources can share a set of standard interfaces and create service description by assimilating definitions of interfaces and resources.

2. **Programming Framework:** For the smooth assimilation of IoT within BP environment, it is necessary that APIs and service artifacts are in place which allows accessing a service provided by IoT devices with the same ease as a method call on an object. A search criterion is specified by the application developer. The application developer also gives the service discovery method which fetches the description file that is available for the nominee service. Next, the developer chooses only those services that are needed to achieve the current situation. Once the processing of service definition files is finished, the next task is creation of a service handler for every IoT service. Then a single service description file is generated which contains data about all services that have been included and service handler provides an API to access the service.

3. **Event Manager:** Supports the interaction, which is event-based, between the components of the framework and this type of interaction in generally known as publish-subscribe systems. The publisher publish the event (structured messages) to the event manager and subscribers can show if they are interested in these events to the event manager via subscriptions. An analogy to explain this can be when users show interest in various channels and subscribe to them via television providers and the people who create the channels are the publishers in this case. An important property of these systems is that it is not necessary for the publisher to know who the subscribers are. This allows the architecture to be adaptable and dynamic.

4. **Device Status Monitor:** The main aim of the device status monitor is to timely report the status of each IoT device within the IoT environment in order to achieve execution of IoT aware BPs that is failure transparent in nature. Device status monitor acts as the publisher and publishes information about probable breakdown of

every device. Both the BP and the service replacement manager are the subscribers and they subscribe to these publications. Device status monitor is like an ECG. Just like the ECG shows that a person is alive by measuring the heartbeat, every device also sends a period message that acts like a heartbeat which shows that the machine is working. If a machine does not send this message within a given time span the device status monitor sends the ID of the device to the event manager which will then alerts the BP as well as service replacement manager to find a replacement for the failed service.

5. **Service Replacement Manager:** As the name suggests the task of this device is to find a replacement service in the face of a glitch. It also needs to make sure the original state of BP is retained so that when service replacement is over BP can be resumed again. The service registry gives a list of candidate services and this is used to find the replacement service it then searches for services that match the definition for the service that needs replacement. Then manager the selects one of the services randomly that match the criterion.

6. **Distributed Business Process:** BPs in this framework need a suitable engine as well as language to design and implement them. The processes are designed by means of business process modeling notations (BPMN) 2.0 and following are the functionalities attained by adopting this language:

   • Communication between the BPs can be defined by domain experts;
   • A large number of events and workflow patterns are supported;
   • A uniform exchange set-up is present;
   • User tasks are provided for the purpose of human communication;
   • By specifying domain-specific elements this language can also be extended thus making future process developments simplified.

Languages that are rule-based like semantic web rule language (SWRL) are combined with BPMN. The rules define the underlying logic for business whereas BPMN takes care of logic for the program. The advantage offered by this is that changing the BP will not need the BPMN code to be changed or accessed. BPMN engine is needed not only on the STB and remote control but also on the smartphone. The domain expert can thereby create an adapted BP for smartphones.

The implementation of this architecture can be broadly divided into three major tasks:

1. Programming abstraction at the application level so the application development process is swift and easy;
2. The STB needs to be designed such that it locally collects, and processes data originated from IoT devices; and
3. Programming of the low powered sensor devices.

Below is a detailed explained how each of these tasks is achieved:

- **Developing RESTful WSs:** The WSs are built on Contiki-based Tmote Sky nodes. Every service acts like an access point for a resource through RESTful API and it also implements generic interfaces. Once the RESTful WSs are set up, they can easily be found at the Internet level. Registry manager allows a REST server to be exposed in the form of a RESTful service. This is accomplished in two stages—automatic generation of service description and resource discovery. In resource discovery procedure, every REST server executes a resource and a record of REST resources is sent back when a GET inquiry is made to this resource.

- **IoT-Aware BP Development:** This step is mainly concerned with the integration of IoT services into the BP. When the application designer wants to incorporate an IoT service into the BP, the first task he performs is searching the services that are available, with the help of an external service discovery protocol to procure WSDL files for available services. Next the WSDLs are processed with the help of a programming framework component corresponding service artifacts are generated.

  Discovered IoT services, available in the service pallet, are then displayed in the integrated development environment. In the end, the task of the application developer is simplified i.e., using a drag and drop facility he can add a number of services into the logic with ease. Any service can be referred to with the help of an API produced by service handler.

- **Event Manager:** The main aim of the event manager is to facilitate the interaction, which is event-based in nature, between the various components of the structure as mentioned above. The implementation is done using publish-subscribe applied to distributed resource scheduling (PARDES) system. The PARDES broker (PB) receives advertisements (request for potential publication from IoT devices) and the BP can subscribe to a specific publication by stating the conditions for subscription.

- **Service Replacement Manager:** This device jumps to action when a particular device is not able to perform its task. The BP is halted and put into wait state and meanwhile the service replacement manager does the following:

  1. **Selection of Nominee Service:** Candidate services are retrieved from the service registry and one service is randomly chosen to take place of the failed service. If alternate services are not found then an exception is generated, and an alert is sent to domain expert who takes care of BP maintenance.

  2. **Actual Service Replacement:** This can be achieved with the help of two methodologies—the whole process is replaced with another instance with newly designed services or only the substituted service is re-binded. Former is accomplished by changing the XML schema of the entire process and then instantiating it at runtime and latter is achieved with the aid of proxy services.

- **Distributed Business Process Execution:** Activiti BPMN execution engine is the BP execution engine to android. A database supporting java database connectivity (JDBC) is needed to store various kinds of data and also uses the XML stream reader. Since these are not available on android, the source code is adapted to Android's XML pull parser and SQLite database. Communication link is established between the engine running on the STB and the BPMN engine on the smartphone via message queuing telemetry transport (MQTT).

The architecture that has been described here has been qualitatively and quantitatively and the achieved results point to the fact that this framework can be successfully employed for future IoT systems and are ideal for integration on IoT in BPs. This REST inspired architecture can act as a foundation platform on which much more advance characteristics and designs can be generated.

9.3.2   BENEFITS OF IoT IN KEY SECTORS

9.3.2.1   MANUFACTURING

Manufacturing is the leader in the Industrial Industry; it is ahead of all industries in the IoT reality. According to IDC data, published in early 2017, money spent in the field of IoT in the manufacturing industry was $178 billion in 2016. The manufacturing sector uses a lot of Machinery which needs to be closely monitored by human surveillance. IoT solves this bottleneck by embedding technology sensors in the machinery which can communicate with other machinery and a head administrator. This creates a network of interconnected machines which are monitored by the computer. At the core of what is, reshaping the manufacturing industry is data.

9.3.2.1.1   The Anatomy of an Informed Manufacturing Plant

An informed manufacturing organization contains four crucial elements which are discussed below:

1. **Products:** Software applications, Technologically Advanced sensors, and controls work together to gather and share real-time data as goods move through the production line. Smart products can use this data to take decisions on their own without human involvement.

2. **People:** People across various business functions can be connected, and provided with useful information in real-time, they can then use this information to provide intelligent design, maintenance, and operations as well as safety and higher quality of service.

3. **Processes:** Laying an emphasis on bi-directional data being shared across the manufacturing value chain from the supplier to the consumer would result in informed processes which would further result in an adaptable and flexible supply chain.

4. **Infrastructure:** Smart infrastructure elements are able to communicate with devices (such as phones, tablets) and through these devices with the people. These smart infrastructure elements are better able to handle the complexities that come up in a manufacturing plant. Manufacturing plants with smart infrastructure are generally more efficient.

9.3.2.1.2 Benefits of IoT in Manufacturing

1. **Smart Factories:** The latest technological advances in IoT has led to the concept of connected machinery which has increased the efficiency of manufacturing plants, where the machines that are being used have the ability of being linked to one another. Networked machines fully automate the process which optimizes production. For instance, a machine can identify a break down which could have been potentially dangerous, shut down other machinery that can be damaged, and directly notify the staff (human involvement) so that they can handle the situation.

2. **Simplified Components:** The physical complexity of the products reduces as the functionality moves from mechanical side to the software side. This shift that happens removes physical components and also eliminates the process needed to build these components. The reduction in the physical complexity results in an increase in the number of sensors and the software which introduces its own complexity, for example: glass cockpits in airplanes.

3. **Reconfigured Assembly Processes:** The field of Manufacturing is evolving in the direction of standardized platforms, with individual products that can be customized later in the entire assembly process. Software can later be uploaded/updated on to the product via cloud; it can be done by the field technician or the consumer himself. Changes in Product design can be included later in the stage also, in-fact even after delivery.

4. **Continuous Product Operations:** For a long time, the process of Manufacturing has been a discrete one that usually ends once the product has been shipped. Smart and inter-connected products can't operate without a properly managed cloud-based technology stack. The stack is also a component of the product that is supposed to be operated by the manufacturer. It must be improved throughout the product's life. In this sense, manufacturing has now become a permanent and continuous process.

9.3.2.2 SUPPLY CHAIN MANAGEMENT (SCM)

The inventory management system relies a lot on guessing, IoT can be used as a tool that brings the advantage of real-time tracking to the inventory.

Lack of real-time tracking would mean that you cannot obtain crucial information such as how much time your drivers are spending active with the load, if they are taking the most effective route, and whether or not improvements could be made to how pallets are flowing throughout the warehouse. Moreover, the IoT also improves inventory counting as Manual data collection is prone to human error. Too many warehouse operators usually spend a rather disproportionate amount of time tracking lost or misplaced pallets as a result of data entry errors. In a connected warehouse, issues like this are eliminated because every single pallet is being tracked throughout the process. IoT devices can track and trace the inventory system of an organization on a global scale. It makes it possible for Industries to monitor their supply chain by obtaining meaningful estimates of the resources available. It includes information regarding the work that is undergoing, equipment collection, and the expected delivery date of required materials. IoT makes the Supply chain management (SCM) more efficient in the following ways:

1. **Asset Tracking:** Numbers and bar codes on the products are tracked and this is the widely known method through which goods are managed throughout the SCM process. Now with the onset of IoT industry in SCM, new RFID and GPS sensors are able to track products from each stage in the industry right to the doorstep of the receiver and sometimes even further. At any point in the lifetime of the supply chain, manufacturers are using these sensors to obtain useful data. This type of data gained from the IoT sensors embedded in the products can help companies maintain a closer check on quality control, product forecasting and on-time deliveries.

2. **Inventory and Forecasting:** With the use of IoT sensors, far more accurate inventory monitoring can be obtained, far more than what humans can manage alone. For instance, Amazon uses WIFI robots which scan QR codes on the products to track the orders. This means that inventory is being tracked, even the supplies that are in stock for future manufacturing just by the clicking a button. This means no deadlines are missed. An added advantage is that the data, which is being generated and collected, can be further used to study the manufacturing scheduled and come up with more efficient solutions.

3. **Connected Fleets:** With the expansion of supply chain, it's very important to make sure that all your carriers, including delivery trucks, shipping containers, suppliers even the vans that go out for

door-to-door distribution are connected to one another. The data generated is an added advantage as manufacturers are using this data to get the right products out to their customers, in time.

4. **Scheduled Maintenance:** IoT that already exists is using smart sensors in manufacturing plants to plan predictive maintenance and prevent down-time that can be costly for the organization.

9.3.2.3 MARKETING

The data generated by IoT enabled devices can be used to collect data about the user's preference which can later be used to show to the user what he/she likes. Some of the ways of using IoT in marketing are:

1. **On-Demand Services are Expected Everywhere:** IoT increases the consumer's expectations from the products in terms of convenience, and if marketing strategy of an organization does not fulfill this expectation, then consumer won't be interested in it and would think that the product is not worth the money.

2. **Smart Marketing and Social Data:** Another way in Social Media is used in IoT is through Social Access Management. These technologies are letting the user interact/connect with real world entities like buildings and other devices using his/her social identity. The field of IoT is advancing and with this advancement, the field of marketing is moving from 'things' just are connected to their users to including social interactions between the user and the things.

3. **The Internet of Things (IoT) Means That Big Marketing Data is Getting Even Bigger:** The relationship that the marketer and consumer share grows with data. Increase in connectivity in-fact results in more data which is smart and highly informative, which is used to come up with more relevant marketing-campaigns which engages more customers, which is the ultimate goal of marketing. In the modern world, technology is deeply embedded in the consumer's daily life, everything that the user does can be tracked and data can be generated corresponding to that, this has increased the scope and depth of the data that can be gathered from the consumer. Marketers are now coming up with new variety of data, which reveal formerly unexplored areas of consumer preferences, which can be used to make the process of marketing even more efficient.

4. **Everything is Marketable:** With the involvement of IoT in the field of marketing, the domain of fields to which marketing can successfully be employed has increased and there's no sector, no service, no product that is unmarketable. In the present world, all of user's behavior, interaction, and social media presence are interconnected and this helps marketers know in great depth about the consumer. They can now study the patterns of consumer habits, get to know what the consumer wants, and accordingly come up with effective marketing strategies to sell their products to the consumer. Recent studies show that IoT in the sector of marketing will support the importance of big data (BD), personalized transactions, mobile/remote marketing, and customer experience.

Marketers analyze the customer's buying habit across the platform, due to which they are obtaining more and more previously unobtainable information regarding the way's consumer interacts with these devices and products and getting a much better insight into the entire process and also in which stage of it the customer is. Real-time interactions and of course targeted (and even fully contextual) ads are benefiting from this, and the consumer is getting personalized suggestions in the form of marketing which is helpful to the consumer, all this is possible with involvement of IoT in the sector of marketing.

9.3.3 BUSINESS MODEL FOR IoT

In order to actually incorporate IoT into businesses, it is also necessary to have a fundamental business model. Everyone is aware of the potential gains that IoT has to offer in business and BA, but people are unsure how to tap this potential. A business model allows businesses to use IoT resources in various business activities and at the same time ensures that this employment of IoT resources leads to profits. Innovation in business models is the new ways to gain advantage in the market over competitors. The model is built on (Gassmann et al., 2014) and consists of four dimensions: Who, What, How, and Why.

1. **Who:** refers to the target market or the target customers.
2. **What:** is the value proposition. It is an assurance of value to be communicated to the customer.

3. **How:** this describes the sequence to deliver the value assurance to the customer.
4. **Why:** captures the fundamental prototype employed to gain value.

The diagrammatic representation of this model is given in Figure 9.4.

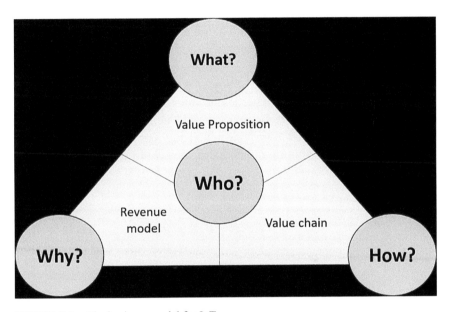

FIGURE 9.4 The business model for IoT.

In IoT ecosystem firms need to collaborate with competitors across industries and thus the traditional business models are not sufficient. Value creation is not based on solving the requirements in a reactive manner; rather it has shifted to solving real-time needs in a manner that is predictive in nature. Value creating IoT can be categorized into three layers:

1. **Manufacturing:** Retailers or manufacturers provide items such as terminal devices and sensors.
2. **Supporting:** Collects data that is used by the layer in which value creation takes place.
3. **Value Creation:** IoT is utilized as a co-creative partner.

IoT architecture can be further divided into four-layer architecture as shown in Figure 9.5. These layers provide different functionalities which in the end create value. With the help of a small example of a health tracker system, these layers can be explained:

- **Gathering Information and Sensing the Objects:** The first step is to collect contextual information. In our example information about heartbeat, blood pressure of the patient is collected.
- **Information Delivering:** Wireless technologies are used to deliver this information to devices that are equipped with the capability of processing this information.
- **Information Processing:** Information which was gathered is now analyzed. For e.g., the heartbeat of a patient is processed and if its value if above/below a particular value, then this is noted.
- **Application and Services:** The results generated in the above layer are now employed to provide actual services to the customer. If the heartbeat was abnormal then alerts are sent to the doctor as well as the patient.

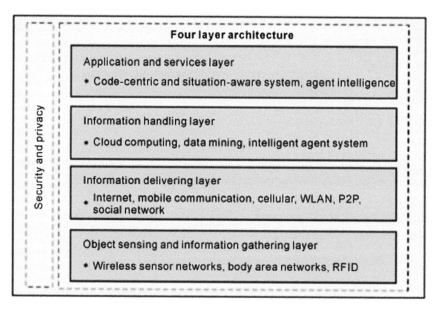

FIGURE 9.5 Architecture of IoT.

The activities of a firm that are involved in value generating tasks include steps from conception to end use. The firm's value chain breaks these down. In the case of IoT, these chains are more complicated, but the fundamental concept is the similar.

There are minimum nine unique classifications along the value chain in IoT in Figure 9.6.

Radios	Chips that provide connectivity based on various radio protocols
Sensors	Chips that can measure various environmental/electrical variables
Microcontrollers	Processors/storage that allow low-cost intelligence on a chip
Modules	Combine radios, sensors, microcontrollers in a single package
Platform Software	Software that activates, monitors, analyzes device network
Application Software	Presents information in usable/analyzable format for end user
Device	Integrates modules with app software into a usable form factor
Airtime	Use of licensed or unlicensed spectrum for communications
Service	Deploying/Managing/Supporting IoT solution

FIGURE 9.6   Layers of IoT value chain.

1. **Strategy:** Process of selecting the appropriate business model for the firm which will allow it to contend with the competitors in the market.
2. **Business Model:** This is the basic fundamental methodology employed by the firm and gives guidelines for operations and value creation methods for customers.
3. **Tactic:** Refers to the choices that need to be made by the firm and but not covered in the business model.

The business framework including these factors is represented in Figure 9.7.

Instead of using the term devices this framework uses the term input because inputs other than devices also exist in IoT architecture. Here network, service, and content have the following meanings:

1. **Network:** With respect to IoT, network basically includes those parts that are concerned with production and manufacturing of product.
2. **Service:** Once production is over the next step is its processing and packaging and all these activities are clubbed under service.
3. **Content:** As the name suggests these includes all the content or in other words the information about the product.

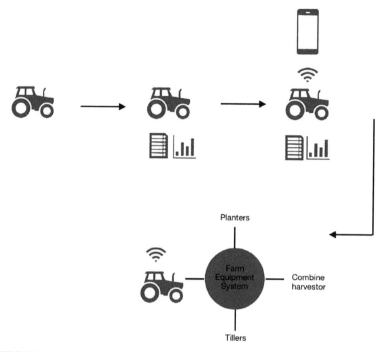

FIGURE 9.7   Proposed business model framework.

Each collaborator's value chain and value proposition are then listed in tabular form. The category of IoT strategy of each collaborator is specified in a column. Further, a column telling about the tactic has been included.

9.3.4   ANALYSIS OF PERFORMANCE AND RELIABILITY OF IoT

As the number of IoT devices increase, the complexity of the infrastructure supporting the end-to-end touchpoints also increases, due to which there is a demand for a highly responsive end to end IoT services, there for measurement of IoT services is essential.

9.3.4.1   AREAS FOR ANALYSIS

Analysis of performance and reliability can be divided into three broad categories and are depicted in Figure 9.8:

- End User and IoT monitoring;
- Application performance management; and
- Infrastructure and database monitoring.

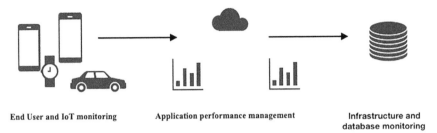

End User and IoT monitoring   Application performance management   Infrastructure and database monitoring

FIGURE 9.8   Different categories for analysis of performance and reliability if IoT.

1. **End-User and IoT Monitoring:** The IoT devices used by the end-user or the applications used by the user to monitor these devices need to be efficient. This comprises of two aspects, the IoT devices, and the application supporting the devices. The individual efficiency, as well as the combined efficiency, drives the performance of IoT at the end user's front.
2. **Application Performance Management:** The data generated by the IoT devices needs to be sent to the vendor server for processing, this involves networking of data from the client side to the server side remotely through the use of the internet. This transfer of data should be as efficient as possible otherwise; it may result in delays which might hamper real time analysis. Cloud services, networking protocols, routing methods drive the performance of IoT in this sub-domain.
3. **Infrastructure and Database Monitoring:** The data being generated by IoT devices has to be stored in a database and later has to be retrieved from this database for processing. This storage and retrieval have to be as efficient as possible.

9.3.4.2 REQUIREMENTS FOR ANALYSIS

Any platform aiming to measure performance and Reliability of an IoT device should make sure of the following:

1. It should be able to monitor IoT applications that use different processor architecture to run (e.g., ARM7, Cortex-M series), and different operating systems (e.g., embedded Linux).
2. It should be able to monitor IoT applications written in multiple languages (e.g., C++, Java, Python, Ruby, JavaScript, Node.js).
3. Overhead of the platform monitoring IoT applications should be as low as possible and needless to say, it should operate within device constraints such as memory, processor power, and network connectivity.
4. It should be able to manage the complexity of software and services offered on the latest IoT device types and applications which keep upgrading with time.
5. It should be able to provide the same user experience, independent of device specifications. In other words, it should be uniform across all IoT devices.
6. It should be able to draw relations between business performance and IoT application performance.
7. It should be able to react to real-time alerts on application or business performance issues. That's what determines its reliability.

9.4   IoT IN AGRICULTURE

By 2050, the population of the world is expected to increase beyond 9 billion people. To meet this growing need agriculture production has to be increased by 50%. Increasing the yield of agriculture is important, and this increase can be brought around by increasing production efficiency. IoT in agriculture gives rise to the field of smart agriculture, which is a term used for IoT solutions used in the field of agriculture, for the benefit of agriculture.

9.4.1   IoT USE CASES IN AGRICULTURE

1. **Monitoring of Climate Conditions:** Weather stations are among the most popular smart agriculture technologies, they combine various smart farming sensors. They collect various sorts of data from the environment and send it to the cloud, which can be later

used to make decisions regarding the type of crop to be grown, or the necessary measures that need to be taken in order to increase productivity. Example-METEO, Pycno, and Smart Elements.

2. **Automated Greenhouse:** Weather stations can simulate a particular type of environment in a closed surrounding. This can be used to help the crops grow irrespective of the natural environmental conditions. Example, Growlink, and Farmapp.

3. **Crop Management:** Another IoT product in the field of agriculture is crop management devices. They are placed in the field for collecting data specific to crop farming; data like temperature, precipitation to leaf water potential, yield percentage, and in general overall crop health. This can help in preventing the crops from any diseases or infestations that may harm your harvest.

4. **Cattle Monitoring and Management:** Similar to crop management there are certain IoT devices that can be attached to animals, to get important information regarding their health and also to monitor the productivity. These devices use smart sensors that are inserted into the animals and are used to keep a check on the animal. They can obtain information about an animal's health, nutrition, productivity, etc. They provide information about the individual animal and also the herd. For example, SCR by Allflex and Cowlar.

5. **End-to-End Farm Management Systems:** This comprises of a number of IoT enabled sensors installed on the farm. These types of systems collect data from all the devices make intelligent reports using this data and also help these sensors communicate with each other. It provides the user with a dashboard to monitor all the sensors at one place. This also enables farming to be done remotely. FarmLogs and Cropio represent such solutions.

6. **Agriculture Drones:** Drones is a form of an unmanned aerial vehicle which is compact in size and can be programmed to do certain tasks or can be manually operated from a remote location. Involvement of drones in the field of agriculture is on the rise. Drones are used for monitoring crops from an aerial view; drones are also used to spray pesticides, irrigation, planting, soil, and field analysis.

7. **Machinery:** Common machines used in agriculture like tractors, harvesters, planters, etc. can be made smart using IoT.

9.4.2   *EVOLUTION OF CONNECTING MACHINES AND FARMS*

The products that are used in the field of agriculture are becoming smart in stages:

1. **Product:** These are the trivial tools that are used in the field of agriculture. Such products help in reducing the manual work of the farmer however, these products themselves are not very intelligent and need human monitoring to be put to use. For example, a tractor is a product that is widely used in farming.

2. **Smart Product:** When these trivial products undergo technological advancements over time, these products become smarter and smarter. Such smart products reduce the human effort and increase efficiency. However, these products act independent of any other products/devices. For example, a smart tractor can be programmed to carry out different tasks according to different situations/climatic conditions.

3. **Smart Connected Product:** These smart products when made part of a network can communicate with one another and exchange data between each other. These products can also be connected to mobile devices/computers and then they can be remotely monitored. For example, the smart tractors mentioned above, when these connect with a phone device, they become smart connected tractors.

4. **Product System:** It refers to a computer program that handles a particular task end to end without user involvement. It connects all the necessary devices/products and makes all of them work in coherence with one another to achieve a common goal. For example, the farm equipment system would handle the task of connecting tractors, harvesters, planters, tillers, and make them work together without human intervention.

5. **Systems of Systems:** This like product system achieves the same goal, but at a higher level. It combines different product systems and makes use of different pre-defined product systems to complete a much broader goal. For example, The Farm Management System would use product systems like Farm equipment system, irrigation system, Weather data system, etc.

Stages 1–4 are shown in Figure 9.9.

Company	Collaborator	Inputs	Network	Service/ processing/ packaging	Content/ Information product	Benefits	Strategy	Tactic
	C1							
ABC	C2							
	C3							

FIGURE 9.9 Different stages in evolution of connecting machines and farms.

9.4.3 CHALLENGES FACED

1. **Hardware:** The hardware refers to the sensor that is being used for recording of the data in the real world. The quality of the sensor used and the restrictions put down on the data being collected due to the limitations of the sensors are crucial when IoT is being applied. The sensors being used should be chosen/manufactured for the agriculture industry in mind.

2. **Analytics:** Also known as the brain of any IoT application should be as advanced and easy to understand as possible. There should be abstraction, wherein the lower layers with high complexity should be hidden and the upper layer which the user would use to interact with the system should be as simple as possible. Data Analytics is important as the collected data itself is of no use, until it is processed and analyzed to extract useful information out of it.

3. **Maintenance:** The hardware sensors being used in the agriculture industry are vulnerable to damage and hence the maintenance of these devices becomes a hassle, therefore to avoid this unnecessary headache and the extra cost of maintenance it is necessary to carefully choose devices that are resistant to conditions regularly found in the field of agriculture.

9.5 CHALLENGES

The introduction of IoT in BA is a relatively new area. The potential growth due to this integration is unprecedented, but at the same time, it is also fraught with difficulties.

The first and foremost challenge is related to the technological aspect. The framework discussed in this platform is still a very basic model and has scope for improvement. More technological advancement is the need of the hour. Not only should the technology advance, but more importantly, it should also be economical. And therein lies the next challenge—monetizing the IoT. These challenges are of three types—the diversity of objects, the immaturity of innovation, and unstructured ecosystems.

9.5.1 DIVERSITY OF OBJECTS

This problem refers to the difficulty of creating business models due to the host of different types of object that come under the umbrella of IoT. The basic idea behind IoT is that everything has an internet presence. Now things can be cars, refrigerators, ovens, and even toothbrushes. This diversity among the objects hinders the ability to create a standardized interface with which they can connect to the internet. Another problem is that decision making is very difficult since there exist numerous ways in which an object, a thing, a business and a consumer can be connected together. Recent estimates show that 99% of the physical objects that can be connected are still not a part of the network and this suggests that there is a high probability that there will be an unprecedented growth in the diversity of objects in the future as more and more objects join the network.

9.5.2 IMMATURITY OF INNOVATION

Many technologies are available today in the field of IoT but the problem with these innovations is that most of them have not yet developed into products and services. They have not yet been modularized or standardized for wider usage and in many cases, engineering work is also needed. Modularized objects that also contain plug and play components are of the utmost importance in the emerging market. Coupling components together allows developers to experiment and generate products and services for an IoT ecosystem and also learn from market experiences while making business models. The innovation must also be advance enough for customers to adopt it rapidly and only then can products become profitable.

9.5.3 UNSTRUCTURED ECOSYSTEMS

The ecosystems that are not properly structured lack a defined fundamental configuration and governance, stakeholder roles and value creation logics.

It is possible that the participants that are necessary for e.g., IoT operators are not present in the ecosystem. Going after new business opportunities means building associations in industries, or strengthening prevailing relationships; it takes time and is difficult for managers. The IoT is still in its early stage and early ecosystems are chaotic and open ground for new participants. There is a dire need for the emergence of ideas that will affect the IoT business ecosystems through innovations in the business model.

9.6   CONCLUSION

The chapter starts by introducing the concept of IoT and also explains the basic architecture. The changing fundamentals of business are the analyzed and the need for IoT in this new era is touched upon. Once it has been established that assimilating IoT in BA is the need of the hour the next step is to act upon this. The action plan is two-fold, i.e., both technical and monetary aspects are a part of it. The technical framework discussed in the framework is REST-based architecture. Since it is equally important to make sure that employment of IoT is adding value to business and also making money the next section covers the ideal uniform business model that can be used for IoT services. Even though IoT offers amazing opportunities its usage is, also fraught with challenges namely—diversity of objects, immaturity of innovation and unstructured ecosystems. The main take away here is that achieving BA through IoT can do wonders for a business and the tools to achieve the same are available. Some improvements and potential solutions to combat the problems are needed, but once these obstacles are overcome, this combination can go a long way in changing the face of BA.

KEYWORDS

- **business analytics**
- **business applications**
- **business model**
- **business processes**
- **data processing**

CHAPTER 10

Business Application Analytics and the Internet of Things: The Connecting Link

ANKITA GUPTA,[1] ANKIT SRIVASTAVA,[1] ROHIT ANAND,[1] and
TINA TOMAŽIČ[2]

[1]Department of ECE, G. B. Pant Engineering College, New Delhi, India,
E-mail: roh_anand@rediffmail.com (R. Anand)

[2]Institute of media communications, University of Maribor, Slovenia

ABSTRACT

The right set of information at the right time is the primary defining factor in decision making, specifically affecting the quality of decisions and their later impacts. It is a known fact that the technological advancement has always marked modification in businesses. With the enormous amount of data being generated, the internet of things (IoT) and big data (BD) have been widely used in businesses to make data-driven decisions to improve customer experience and generate revenue. This chapter discusses business analytics (BA) Applications through IoT, capturing the real-time data production, management, and analysis. The eventual objective of this framework is to give a comprehensive and detailed knowledge about the varied domains of IoT, BD and BA applications. The chapter further outlines the various advantages and disadvantages of the shift in business models. It further takes into account the effect of social IoT on the business process and a case study for the better understanding.

10.1 INTRODUCTION

One of the most important sectors that experience all-time advancement is the Business sector. Moreover, it also plays a determining the influential role in any economy. Business exactly is a blend of human and technological mold in a perfect and impactful manner to yield the maximum useful and definitive output (Kang and Choo, 2018; Gierej, 2017; Steenstrup and Gartner, 2015; Benazzouz et al., 2014).

The internet in recent times has taken over everything from homes to education to health to agriculture to business and this has been made possible due to the vast and fast-growing use of the internet of things (IoT) in every domain possible. IoT allows the connection between people and things to make a smart environment for transmitting the data over the network. In broad concept, IoT can be defined as "a network of dedicated physical items that contain embedded technology to sense or interact with their internal state or the external environment" (Steenstrup and Gartner, 2015). In short, IoT is not just the IoT, but it is the Internet of Everything with limitless applications. IoT has allowed automated control of everything from storage units to warehouses enabling inventory tracking and management. The data collected using IoT devices and BD analytics can be shared and interpreted for the benefit of the business and the customers adding more productivity and efficiency. IoT devices with the huge amount of collected data form patterns to represent how the consumer interacts with the available market. With better and analyzed information about the customer and the market, businesses can reach new domains of success. The fundamental concept of any IoT setup in business is to extract more output in lesser time. Moreover, IoT devices introduced the concept of remote working to companies. Now, physical presence is not mandatory to do a job or complete any task. It allows workers to connect and work at their place of convenience; further results show that the workers working from home or their place are happier and ultimately prove to be more productive for the firm. IoT in business is an example of quality in quantity.

At the same time, it has some issues dealing with its technology, the business model and the various social and environmental impacts that need to be tackled to reach IoT's full potential (Benazzouz et al., 2014). More or less, businesses that are accepting this technological change and modifying accordingly plainly see a higher growth rate than any other firm following the typical traditional methods of decision making.

10.2   THE INDUSTRY CHAIN

Industrialization and Globalization have marked various leaps in technologies concerning the advent of better return outcomes, keeping market scenario in central perspective at the same time, making things easier and beneficial for many generations to come. This has also led to the heterogeneous Industrial Revolutions. The first among them was Industry 1.0 in 1784 followed by Industry 2.0 in 1870, further leading to the development of information technology (IT). These changes have modified the economy and BA. Finally, there came the age of Cyber-Physical systems, the Industry 4.0 and now the world is moving drastically towards Industry 5.0, the age of human and smart system collaboration.

10.2.1   IoT AND EVOLVING BUSINESS MODELS

Table 10.1 shows a yearly based table to show the historical trend of IoT in BA. Whether the traditional model or the modern IoT based model, every business model is governed by some parameters such as: revenue streams, customers' relations, value prepositions, real-time work management, cost-effectiveness, and many others.

10.3   IoT AND BUSINESS: THE CONNECTING LINK

10.3.1   THE INTERNET OF THINGS (IoT)

The IoT is that technology that is influencing every domain possible from daily life, to business to even the economy as a whole. Since the world is making advances at a very rapid rate, BA are bound to increase and IoT has allowed making an impactful connection between people and things at any time or place with the help of individual networking devices. It is merely making devices talk to each other on a simple and easy note. These include everything ranging from smart systems to wearables, mobile phones, healthcare equipment, security devices to smart homes and smart cities.

Many events and procedures are still in progress and many have become realities to redefine IoT concerning BA and world-changing business models. This also helps in establishing a network between endpoint

TABLE 10.1 Year Wise Business Model Representation

Year	Model	Key Elements	About	References
2010	Object Automation Model	Software, Information, Customer Report	Physical objects were digitized using the various technologies to establish digitalized gain characteristics.	Yoo et al., 2010
2011	Business Canvas Model	Subscription Fees, Usage Fee, Software, Information, Customer Resource	A value proposition and customer perspective based business model by taking the importance of information into consideration	Bucherer and Uckelmann, 2011
2012	DNA Model	Software, Information, Customer Resource, IT Cost, Infrastructure, Software, Information, Customer Resource	A business model based on Design, Needs, and Aspirations. This model is used to build a strategic approach towards business and management.	Sun et al., 2012
2013	Multiple open platform (MOP) model M2M model	Performance, Customization, Share, Convenience IT Cost, Infrastructure	A multidimensional structure consisting of various dimensions such as technology, industry, policy, and strategy The machine-to-machine connections increased drastically and reached 195 million	Li and Xu, 2013
2015	Business canvas model	IT cost, infrastructure, convenience, customization, Integration Ability, Software Developer, Data Analyst	Looking at Business models with the block perspective. Every model was now thought of as a mixture of several building blocks compiled together	Dijkman et al., 2015
2017	Stage-Gate system model	New innovation approaches, software, information	An IoT based business model aiming at shortening technological cycles, thus creating new innovation approaches.	Tesch et al., 2017
2018	Service Business model	Manufacturing firms, software, Customer Resource, Management, Platform & Resource Integration Ability	This is a six parameter guiding business model for manufacturing firms based on the application of IoT offering new opportunities of design, technical, and ecosystem lens	Lai et al., 2018

devices for system reliability and focused management (Leminen et al., 2012; Xiaocong and Jidong, 2010).

The concept of IoT in Business dates back to 1999 when British entrepreneur Kevin Ashton used it to name the communication system of the material world with computers by using sensors in which objects had direct or indirect data collection, processing or exchange attributes (Ashton, 2009). The BA foundation saw a significant rise in 2008 and 2009 when the number of devices connected on the internet exceeded 6 billion. CISCO Systems Inc. called it the Internet of Everything. So, this was how IoT came into the business scenario. The introduction of IoT in the business sector as an attribute has its pros and cons. On the one hand, it acts as a tool for connecting people, connecting devices, making them smart for technological and economic growth with the proper transmission and handling of data over the internet, but on the other hand, there exist technological pull forces where existing areas are analyzed for the benefits by the widespread deployment of IoT.

A graph representing the number of installed devices in the recent years and number of expected installed devices in the upcoming years is shown in Figure 10.1.

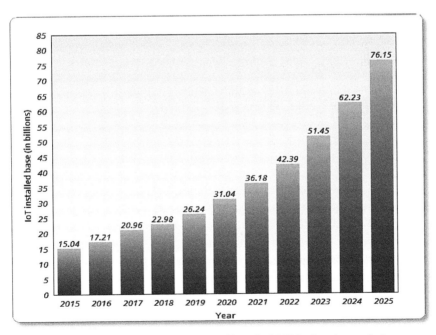

FIGURE 10.1 Graph showing the number of installed IoT devices (year wise).

IoT has facilitated global networking by connecting a vast and varied amount of devices, people, and goods all over the world, thus grasping both the providers' and the users' attention in a single framework (Yerpude and Singhal, 2017).

10.3.2 ROLE OF BIG DATA (BD)

Specifically, BD is the connecting link between IoT and BA. The use of BD arises due to the need for management, transmission, operation, storage, analysis, and visualization of a large amount of data over the Internet. It is a reflection of the changing world we live in. The more things change, the more the data is generated and recorded for future use. So, here it works on a simple mechanism. Each body is given a unique and different accessory to collect data as much as possible without any ambiguity using a network. All this data collection can be further processed to reach useful results, hence marking the exciting trends based on advanced analytics models. This process of churning and management of massive data implementing avant-garde analytics technique to dredge-up hidden patterns and correlations is termed as BA (Sagiroglu and Sinanc, 2013).

10.3.2.1 FACTORS

The three main deciding factors of BD are 3Vs:

- Variety;
- Velocity; and
- Volume.

Variety means a wide variety of data are supported which can be even interchanged into the required format. It also refers to the enormous variety of ordered and disordered data such as texts, tweets, pictures, emails, and many more. By velocity, it means it's capability of handling all types of data whether the sensor data or stock data or large volumes of data. The same is shown with the help of Figure 10.2.

Data is growing at a million rates causing nuclear data explosion. BD comes to the rescue of enterprises against this nuclear data explosion, making incoming of data faster and reducing processing time to prevent blockages. Volume refers to the management and processing of a massive amount of data. It encompasses the data available to be assessed for relevance. There are millions of people sharing data and with increased amounts of IoT devices, data generation is on a high rate, resulting in the large amount to volume in BD analytics.

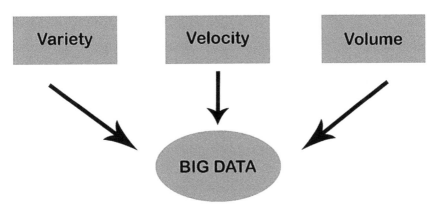

FIGURE 10.2   The 3Vs of big data.

10.3.2.2   BENEFITS FOR BUSINESS ORGANIZATIONS

A click away from self-service-BD has made it possible for companies to expand their services and products with just a click of a button without involving any human intervention:

1. **Resource Pooling:** Using a multi-tenant model, resources like storage, memory, virtual machines (VMs) can be grouped for efficient usage.
2. **Availability of Data:** Information is available all time over the network and can be accessed with various devices simultaneously.
3. **Elasticity:** Resource elasticity means resources can be increased or decreased as per customer's need.

The use of BD is expanding at a rapid rate in all domains of life including engineering, biological, and biomedical domains. In the purview of the growth in the number of connected devices using IoT, this led to an increased amount of data to manage, that completely reflects how and why BD overlaps with that of IoT. The concept of BD completely nullified the need for transactional data produced manually. Instead, BD with the IoT networks unfolding all over the world gave rise to a new feature of sensor data in which data is generated and collected by the connected IoT devices (Barnett, 2015).

10.3.3   BUSINESS ANALYTICS (BA)

BA mainly refers to technologies, applications, and practices for the collection, processing, and analysis of data to yield fruitful outcomes and working with these outcomes to ensure better customer results (Balachandran and Prasad, 2017). The primary purpose of BA is to ensure fast and reliable decision making. This is highly required in any business domain to be successful and to land on marks amidst the competing world outside.

No set of humans can collect, store, and analyze such a vast amount of data present today in the world; then there comes the role of BD that helps business models to deal with an enormous amount of structured, semi-structured, and unstructured data. This evolving and meaningful data analyses, data sets to uncover unknown patterns and insights. The comprehensive goal of BA is to extract useful information from all data available to transform into a coherent structure for further use. The processing of data implementing the advanced techniques in the Business domain to derive and predict useful decisions is called BA (Yerpude and Singhal, 2017).

BA has been further divided into descriptive, predictive, and prescriptive analytics as shown in Figure 10.3. A closer look shows that:

1. **Descriptive Analytics (DESCBA):** This uses data mining, intelligence, and web analytics to present the trending information of the near past and current events, helping business models to know what are the demands and drawbacks of the market.
2. **Predictive Analytics (PREDBA):** This uses statistics and network analytics to forecast future models.
3. **Prescriptive Analytics (PRESBA):** This uses AI, optimization, and reasoning to provide the most suitable set of models for the organization to choose from indicating the pros and cons of each.

These layers also contain two interconnected analytics: Inquisitive Analytics and Preemptive Analytics. The inquisitive analytics uses statistical and factor analysis to approve/reject business prepositions while the preemptive analytics concerns with the precautionary actions on adverse events, providing alternative and sanative strategies to cope up.

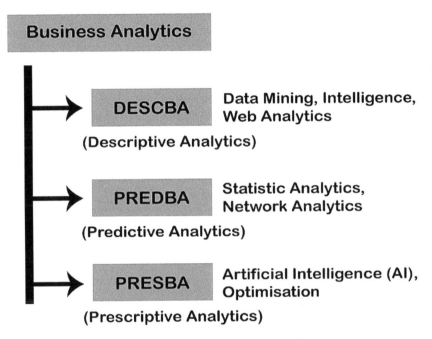

FIGURE 10.3   Different types of business application analytics.

10.4   HOW IS IoT ANALYTICS DIFFERENT?

It has been predicted that by 2025, more than half of the BP and systems would be digitized, utilizing the benefits of IoT and BA applications.

Analytics typically means reporting, analyzing, and predicting the results, while on the other hand, IoT requires Analytics for the collection, analysis, and filtration of data to reach boastful results for BP.

10.5   BUSINESS APPLICATION ANALYTICS THROUGH IoT

The term BA has evolved over the years with different technological growth to witness a market boom. It has a long history of functions, operations, and applications re-defined through ages to help modernize the market with the tincture of technology in them (Duan et al., 2018). Technological growth is the determining factor in shaping business growth and providing the original stimulus for business modernization through enriching business operations, better customer services and upgrading products.

Figure 10.4 explains the insights of IoT and BA in a detailed view. It shows how the overall combination of IoT, BD, and BA is required to aim and attain competitive objectives like volume flexibility, low cost, quality, speed, production customization, and dependability. This further leads to the business performance growth in the form of net profit, market share and a strong customer base building.

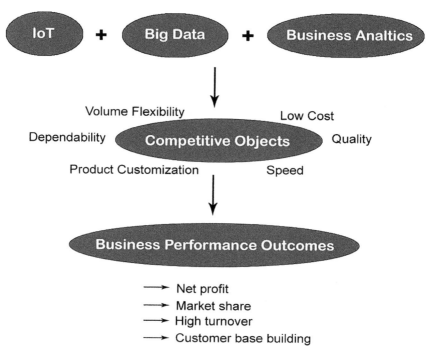

FIGURE 10.4   The connecting link between IoT and business analytics.

10.5.1   *ANALYTICS AND OPPORTUNITIES*

BA applications range from web analytics to customer analytics. BA has evolved from Analytics 1.0 dealing with business intelligence to Analytics

2.0-the era of BD and is advancing towards Analytics 3.0 dealing with data embellished offerings.

IoT and Analytics add icing on the cake by reinforcing the quality of business applications in all domains possible till now. Using IoT and BD analytics together opens ways to new forms of business. The development of IoT and its applications for BA generates higher economic value in two approaches:

1. Opportunities for transformation process; and
2. Opportunities for new business models (Del, 2016).

IoT fosters innovations in business firms, bringing the whole new interesting scenario with machines, no matter electrical or mechanical or a device with or without 'digital' nature to be connected to behave in an intelligent manner analyzing results, taking into account the drawbacks and then showing the most suitable outcome. IoT has played the vital role of acting as one of the pillars of Business process management these days with the increased productivity and reduced marginal costs (Del, 2016). Moreover, Analytics is not a technology of itself; it requires a perfect combination of various tools involved in the collection of information, its analysis and prediction of the various results (Stankovic, 2014). Data integration and data mining are the foundations of advanced analytics. It further allows "360 views" of operations and hence helps in a proper analysis of data to reach beneficial outcomes (Bose, 2009).

The introduction of IoT in the Business domain has developed various research areas such as statistics and machine learning (ML). The results hence obtained are used to open, advance, and automate decision-making and building a firm knowledge foundation for the upcoming future (Yerpude and Singhal, 2017).

For example: SnapSkan is a free service that helps customer aware of their tyre's conditions and their impact on road safety. There are scanners installed in garages and parking lots that read data and store it to provide you when one asks for.

Through time, more, and more business models are becoming IoT friendly, this has broken the old rule of monotony because IoT requires companies to communicate with one another for data to flow between them thus disabling the traditional business models for IoT (Chan, 2015).

Directly connecting devices to the internet and collecting information is not enough. Using IoT for BA applications demands the proper knowledge of how, when, where, and what data to process to reach a useful result. IoT eliminates the need to look into the past; instead, it makes predictions and looks directly into the future by using real-time information.

Some facts and figures advocating rapid and increased use of BA applications using IoT are:

By 2025, it is predicted to have:

- More than 70 billion connected devices;
- 18–20% data from IoT;
- More than 3 million increase in income for digitally transformed business on an average.

10.5.2 ARCHITECTURE

The BA application through IoT involves mainly three layers shown in Figure 10.5:

- Perception layer;
- Network layer; and
- Application layer (Ju et al., 2016).

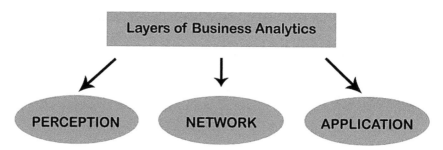

FIGURE 10.5   Layers of business analytics.

The perception layer is the base layer of BA architecture. Its primary function is to collect and organize data from different sources with the help of sensors and RFID tags in the local network. The network layer transmits this collected information for real-time utilization. In this layer, information is not only exchanged between people and things but also among things and the various devices involved on their own. Next comes the application layer, which is the combination of analysis and decision making to reach the satisfactory results to fulfill the customers' demand.

The combination of three layers is essential for the proper functioning and application of BA architecture.

10.5.3 EXAMPLES

IoT technologies are being implemented in different organizations and industries to the significant effect. Some of them are discussed as follows:

1. **Advantech:** It is a leading firm in IoT intelligent systems helping restaurants these days to use integrated systems with the digitized menu and smarter point of sale (POS) devices along with the sensors to collect information to embellish processes, reduce cost, and provide better services.

2. **Whirlpool:** As any electronic company, Whirlpool also wants to know the machine's as well as the customer's feedback. For this, they have installed sensors on their washing machines and refrigerators to gather data about the usage of product and its response to it.

3. **Ibaco Ice Cream:** This is a Bombay-Chennai based ice cream chain that faced problems due to melting away of ice cream due to improper storage. Nimble Wireless, an IoT hardware company installed the temperature sensors that allowed the remote monitoring of an ice-cream store.

4. **Chai Point:** This is a tea serving franchise that uses an integrated IoT and AI tea dispensing machine with a dashboard to track your tea through machine, thus providing you the controls to alter milk and water levels.

5. **Texa:** It is another IoT technology installed in cars to get in-depth information about a vehicle's metrics. This includes everything from fuel consumption to a driver's skill and engine health. This technology is so smart that it sends notification in the case of an accident.

6. **Steelcase:** Lack of fully optimized workplace is yet another demanding problem that companies face today merely because of the low employee engagement. Steelcase is an IoT technology helping the companies to cope up with this. Using this, any employee can send information about any part of the office that is not in use at any particular time. Steelcase then further processes this information to find ways about how to optimize the available area, thus helping to increase the adaptability across the firm.

learn about the wide spreading and common health problems of recent times. Through a survey, it has been found that more than half of physicians use social media for professional reasons (Lee, 2015).

3. **Specialized and Distinct Communities:** Social media has the most significant asset that they classify people according to their interests. The business companies can take advantage of this factor and market their product and send notifications and recommendations according to the person's interests.

4. **Competitive Atmosphere:** With the advent of IoT, new business domains are bound to emerge. This would lead to a competitive atmosphere in the business domain to gain benefit from the digital ecosystem.

This concludes that a combo of people with services and digital media enormously brings business benefits to extreme extents.

10.9   INTEGRATING THE BUSINESS MODELS AND STRATEGIES WITH IoT

The business sector with IoT is emerging at a very rapid rate. The business models that have already evolved themselves with IoT are bound to grow and gain profit early. Shifting to IoT though demands challenging changes and difficulties, but the use of the right IoT technology for business would act as the defining line between success and filth.

Any business model can change and modify for its betterment using IoT. Some of the live examples are:

- 'NIKE' has started the technology industry with their watches and mobile apps.
- 'John Deere' is yet another equipment selling company, which is now also providing data-backed information for the farmers' benefit (Gubbi et al., 2013).

10.10   DISADVANTAGES

It is one of the well-known facts these days that IoT helps companies to be more efficient in their process of decision making and provide better customer services, but it has some disadvantages too. Some of them are lack of standards, environmental concerns and its high cost. The primary aim of IoT is to develop a cross-industry technology; instead, different IoT standards are competing among them, making it highly competitive. A few significant drawbacks of IoT are discussed below:

1. **Data Acquisition and Warehousing:** This challenge is about the collection and storage of data for decision making purpose. The principal barrier to the analysis of BD arises due to lack of data precision, data accountancy and divergence of scale built-in data collection and storage further generating speed and resolution problems (Wang and Wiebe, 2016).

2. **Data Mining and Cleansing:** This challenge arises due to the proper extraction and cleaning of data to be used for various purposes. Due to the vibrant, diverse, and interrelated connection between data, it becomes very difficult to clean data and then process information (Chen, 2013).

3. **Data Aggregation and Integration:** This challenge deals with the integration of data from various sources after the cleansing of data to reach useful results. The availability of data in a huge amount and its diverse nature remains a significant difficulty towards the smart integration of data (Edwards and Fenwick, 2016).

4. **Privacy:** It is yet another throwing challenge in BA using IoT. With such a large amount of personal as well as professional data of individuals being available today, privacy concerns are high. Such availability of location-based-information to private organizations and non-elementary bodies of society poses severe threats to customers (Krishnamurthy and Desouza, 2014).

5. **Data Governance:** With the rapidly increasing amount of data, the concern to extract the right type of data for a particular workfield is necessary; many organizations are facing this problem of the collection of right data at the right time to attain insights in business decisions and various operations (Hashem, 2015).

10.10.1 PRECAUTIONS

Certain ways can help out the organizations to minimize their risk of IoT security:

- **Review the Risk:** It refers to penetration techniques to assess the risk of connected devices, evaluate, and study the risk, thus maintaining the priority list for being known to the primary security concern.
- **Encryption of Data:** Within all connected devices, end to—end encryption is established, making sure that your data is safe in rest or transit stage.
- **Authentication:** Refers to re-review of connections made to your device and taking care that the critical and real connections are authenticated.
- **Integration:** It includes secure booting every time the device starts up, securing updates and the codes being run on the device with the help of code integrity to ensure integration.
- **Strategical Scale:** It means that one should have a scalable security framework and architecture which supports the IoT deployments.

10.11 A CASE STUDY: A REAL WORLD EXAMPLE OF IoT IN BA ANALYTICS

Name: Altizon
Year founded: 2013
Headquarters: Pune, India

1. **Problem:** There are many industries where machines are connected for various processes but still there is a need for human intervention, i.e., their connections are not yet fully automated. Lack of real-time visibility into machines and operations reduces drastically manufacturers' ability to make data-driven, better decisions, and change their old traditional method of predictive analysis to real-time analysis and decision making.
2. **Solution:** Altizon IoT platform, named 'Datonis' helped companies to remodel their firms using BD, ML, sensors, and RFIDs for the overall benefit of the business. Datonis consists of three components: Edge connectors, back end platform and Intelligence component.
3. **Edge Connectors:** It is used for a variety of purposes. They act as connectivity protocols between devices and finally connect them to the manufacturers' network. Edge is built in with cache memory for storage of data when offline. The primary function of Edge

is to process and analyze data present on edge, thus enabling the required and essential data to transfer to the platform.

4. **Back End Platforms:** This integrates the connected devices, connecting physical devices with IoT software. It uses BD and ML to help businesses get real-time access to data for better and economic decision making, a proper use of assets and improved customer services, ensuring the generation of new revenue and visible growth.

5. **Manufacturing Intelligence:** It provides the overall view of manufacturing operations data.

These all tools helped Altizon to reach its present extent because of unified data view and real-time decisions.

10.12 HIGHLY MISINTERPRETED THINGS ABOUT IoT IN BA APPLICATIONS

With the advent of IoT in BA applications, much of the focus is given to technology while the primary goal of any firm or business startup has taken a backseat. Business is all about structured planning and appropriate budgets; it's not about technology, but about the most efficient use of technology to cut costs, increase revenue and enhance customer services. IoT based enterprises must not be distracted by the fantasizing face of technology but should remain focused on the main aim of technological use in their areas of work to achieve what they want. Some of the common misconceptions about IoT in BA applications are as follows:

1. **Internet of Things (IoT) Is About Information, Not Devices:** The value of IoT for any firm resides in the data that IoT devices collect. Data collected using sensors and other devices can be used to know the current state of business, its strengths, and weaknesses, and faster, more informed decisions. With the use of artificial intelligence (AI) machines and analytics, new data gathered can be analyzed opening new business domains and ways of revenue generation.

2. **IoT-Business Sensors:** IoT is not all about sensors, the use of IoT in an impactful manner in any business domain demands use of other connected and well-versed devices such as printers, phones, other mechanical machines and many more. Connecting IoT with

sensors would be an injustice to all the other devices connected which need to be connected for reaching the output.

3. **Foolproof Security Planning:** With such vast amount of data, there is always a risk of security. Enterprises must take proper precautions from the very first day of their business. Proactive cybersecurity and privacy programs must be developed to nullify the maximum risk rate possible.

4. **IoT Projects Are Advanced and Excellent:** More than 70% of the IoT based business setups fail, due to lack of planning in initial stages. People misinterpret IoT; they forget that IoT connections are just means for a well-defined and useful decision-making process. What kind of decisions one interprets by data analysis ultimately depends on business executives.

5. **IoT and Efficiency:** IoT is often linked with efficiency but benefits of IoT clear business insight, better customer relations, competitiveness, sustainability of a firm and many more.

6. **IoT Alone Is Sufficient:** This is not true, with an increasing number of connected devices; the amount of data is bound to grow exponentially. Then there comes a need of AI, to interpret useful results. Businesses need to possess a sound knowledge of both the technologies to grow and flourish.

The transformation of Business models with IoT is undeniable. With more access to data and better insights, better decisions can be made. However, the installation of IoT devices and their connections must be smart enough to gain success. Every IoT installation might not succeed. IoT projects are complicated and need to be accounted for structured planning. It is said that for accelerating the IoT implementation; organizations should balance gaining efficiency and maintaining security and privacy. Businesses can achieve new heights with the proper implementation of IoT. Instead of taking IoT as a technical buzz, businesses should aim to solve a problem (i.e., the main focus must be the problem) by utilizing IoT as a boon.

10.13 CONCLUSION

With Industry 4.0, IoT has gained an impactful position in all domains. IoT allows devices to connect anytime and anywhere using sensors and RFIDs. This gives birth to the Smart environment, consisting of smart devices connected to the transmission of information. This collected information is then processed and analyzed to obtain useful results. The process of Business Decision making has highly evolved with the use of IoT. IoT has enabled online BA to increase rapidly. Data is fed for analysis to find hidden customers' interest patterns and make adequate plans accordingly while the current data is used to check the relevancy of the present business model.

BA is a vast and varied growing region to work on. Using IoT and Analytics shows firms that the right decision making at the right time and place is the key to perfect successful business in today's smart and evolutionary growing environment. Study of current states of IoT though marks some flaws of insufficient information and security issues revealing many research directions. Role of IoT and Analytics in Business Process Management is yet to unfold.

KEYWORDS

- **automated work management**
- **big data**
- **business analytics**
- **collaborative networking**
- **IoT analytics**
- **real-time management**

REFERENCES

Al-Mashari, M., & Zairi, M., (2000). Revisiting BPR: A holistic review of practice and development. *Business Process Management Journal, 6*(1), 10–42.

Ashton, K., (2009). That 'internet of things' thing. *RFID Journal, 22*(7), 97–114.

Balachandran, B. M., & Prasad, S., (2017). Challenges and benefits of deploying big data analytics in the cloud for business intelligence. *Procedia Computer Science, 112,* 1112–1122.

Barnett, G., (2015). *Harnessing Data in the Internet of Things: Strategies for Managing Data in a Connected World.*

Benazzouz, Y., Munilla, C., Gunalp, O., Gallissot, M., & Gurgen, L., (2014). Sharing user IoT devices in the cloud. In: *Internet of Things (WF-IoT), IEEE World Forum* (pp. 373, 374). IEEE.

Bose, R., (2009). Advanced analytics: Opportunities and challenges. *Industrial Management and Data Systems, 109*(2), 155–172.

Bucherer, E., & Uckelmann, D., (2011). Business models for the internet of things. In: *Architecting the Internet of Things* (pp. 253–277). Springer, Berlin, Heidelberg.

Chan, H. C., (2015). Internet of things business models. *Journal of Service Science and Management, 8*(04), 552.

Chen, J., Chen, Y., Du, X., Li, C., Lu, J., Zhao, S., & Zhou, X., (2013). Big data challenge: A data management perspective. *Frontiers of Computer Science, 7*(2), 157–164.

Del Giudice, M., (2016). Discovering the internet of things (IoT): Technology and business process management, inside and outside the innovative firms. *Business Process Management Journal, 22*(2).

Dijkman, R. M., Sprenkels, B., Peeters, T., & Janssen, A., (2015). Business models for the internet of things. *International Journal of Information Management, 35*(6), 672–678.

Duan, Y., Cao, G., & Edwards, J. S., (2018). Understanding the impact of business analytics on innovation. *European Journal of Operational Research.*

Edwards, R., & Fenwick, T., (2016). Digital analytics in professional work and learning. *Studies in Continuing Education, 38*(2), 213–227.

Gierej, S., (2017). The framework of business model in the context of industrial internet of things. *Procedia Engineering, 182*, 206–212.

Gnawali, O., Moss, D., Clark, R., Jones, B., Eason, W., & Shirkalin, D., (2016). Scaling IoT device APIs and analytics: Information processing in sensor networks (IPSN). *Proceeding of the 15th ACM/IEEE International Conference*, 11–14.

Gubbi, J., Buyya, R., Marusic, S., & Palaniswami, M., (2013). Internet of things (IoT): A vision, architectural elements, and future directions. *Future Generation Computer Systems, 29*(7), 1645–1660.

Hashem, I. A. T., Yaqoob, I., Anuar, N. B., Mokhtar, S., Gani, A., & Khan, S. U., (2015). The rise of "big data" on cloud computing: Review and open research issues. *Information Systems, 47*, 98–115.

https://scrape.works/blog/top-7-applications-of-iot-in-business/ (accessed on 16 February 2020).

Ju, J., Kim, M. S., & Ahn, J. H., (2016). Prototype business models for IoT service. *Procedia Computer Science, 91*, 882–890.

Kang, B., & Choo, H., (2018). An experimental study of a reliable IoT gateway. *ICT Express, 4*(3), 130–133.

Krishnamoorthi, S., & Mathew, S. K., (2018). Business analytics and business value: A comparative case study. *Information and Management, 55*(5), 643–666.

Krishnamurthy, R., & Desouza, K. C., (2014). Big data analytics: The case of the social security administration. *Information Polity, 19*(3/4), 165–178.

Lai, C. T. A., Jackson, P. R., & Jiang, W., (2018). Designing service business models for the internet of things: Aspects from manufacturing firms. *American Journal of Management Science and Engineering, 3*(2), 7–22.

Lee, H. J., (2015). A study on social issue solutions using the "internet of things" (focusing on a crime prevention camera system). *International Journal of Distributed Sensor Networks, 11*(9), 747593.

Leminen, S., Westerlund, M., Rajahonka, M., & Siuruainen, R., (2012). Towards IoT ecosystems and business models. In: *Internet of Things, Smart Spaces, and Next Generation Networking* (pp. 15–26). Springer, Berlin, Heidelberg.

Li, H., & Xu, Z. Z., (2013). Research on business model of internet of things based on MOP. In: *International Asia Conference on Industrial Engineering and Management Innovation (IEMI2012) Proceedings* (pp. 1131–1138). Springer, Berlin, Heidelberg.

O'Leary, D. E., (2013). Big data,' the 'internet of things' and the 'internet of signs. *Intelligent Systems in Accounting, Finance and Management, 20*(1), 53–65.

Sagiroglu, S., & Sinanc, D., (2013). Big data: A review. In: *Collaboration Technologies and Systems (CTS), International Conference* (pp. 42–47). IEEE.

Stankovic, J. A., (2014). Research directions for the internet of things. *IEEE Internet of Things Journal, 1*(1), 3–9.

Steenstrup, K., & Gartner, K. D., (2015). *Inc.: The Internet of Things Revolution: Impact on Operational Technology Ecosystems*. Gartner Research Note 3.

Sun, Y., Yan, H., Lu, C., Bie, R., & Thomas, P., (2012). A holistic approach to visualizing business models for the internet of things. *Communications in Mobile Computing, 1*(1), 4.

Tesch, J. F., Brillinger, A. S., & Bilgeri, D., (2017). Internet of things business model innovation and the stage-gate process: An exploratory analysis. *International Journal of Innovation Management, 21*(05), 1740002.

Wang, Y., & Wiebe, V. J., (2016). Big data analytics on the characteristic equilibrium of collective opinions in social networks. In: *Big Data: Concepts, Methodologies, Tools, and Applications* (pp. 1403–1420). IGI Global.

Xiaocong, Q., & Jidong, Z., (2010). Study on the structure of "internet of things (IoT)" business operation support platform. In: *Communication Technology (ICCT), 12ᵗʰ IEEE International Conference* (pp. 1068–1071). IEEE.

Yerpude, S., & Singhal, T. K., (2017). Internet of things and its impact on business analytics. *Indian Journal of Science and Technology, 10*(5).

Yoo, Y., Lyytinen, K. J., Boland, R. J., & Berente, N., (2010). *The Next Wave of Digital Innovation: Opportunities and Challenges: A Report on the Research Workshop 'Digital Challenges in Innovation Research.'* Available at SSRN 1622170.

CHAPTER 11

An Enablement Platform for an Internet of Things Application: A Business Model

B. D. DEEBAK

School of Computer Science and Engineering, Vellore Institute of Technology, Vellore–632007, India, E-mail: deebak.bd@vit.ac.in

ABSTRACT

Internet of things (IoT) has advanced the evolution of the Internet that connects the physical objects to obtain the Internet access in order to communicate and sense the physical environment with other devices. A significant phenomenon is to ramify the objective of business model that has a lack of literature studies to appraise the scope of business models, i.e., how does various IoT application transform into design creation, model delivery and value captivation. Therefore, this chapter retrospects the design strategies of business model and addresses the researches gap to analyze the IoT application in order to enable the business platform models. Moreover, this business platform plays a crucial role to connect the IoT devices and the enterprise IoT applications as distributed business application software. The business IoT model is preferably chosen to enable qualitative based interview approaches that allow the elemental business models to discover. In general, the elemental model inquires the business model objectives and their relevance that construct the elemental blocks to determine whether the previous studies research on IoT based business model for generic application systems or not. Moreover, the comparative studies show that how the business IoT application models differ the generic IoT applications.

This chapter chiefly focuses on the theoretical implication that is related to the previous literature studies on IoT and other functioning phenomena. It shows the resultant studies in terms of business model, building blocks, and its primitive types, which are the key factors of business IoT model. Notably, this chapter thoroughly reviews the business IoT modeling platform with generic IoT applications to determine: (1) Importance of value proposition and customer relationships to construct the business building blocks; and (2) Dissimilarities and IoT practitioners with real-time application to refer the business findings.

11.1 INTRODUCTION

This chapter deals with the business model of internet of things (IoT) that is mainly focused on the modeling platform of IoT. Nowadays, IoT is transforming its facets into several distinct features that connect the physical objects with Internet to obtain and communicate the sensing information between the real time devices (Mishra et al., 2016; Borgia, 2014). As IoT has remarkably evolved for the progress of next-generation networks, a significant change is expected to provide a world-class business objective, i.e., for the people's day-to-day activities. A management consulting firm known as Mckinsey & Company reports that the revenue generated by IoT would be approximately equivalent to 11% of world's economy by 2025 (Mckinsey & Company). Similarly, CISCO, and General Electric announce that IoT growth would expect to be 50 to 20 trillion US dollars in the outgrowth of world's economy by 2020 (CISCO and General Electric). The above analysis shows that the AI convergence, machine learning (ML), Rich Context and Real-Time Streaming provided by IoT sensor and networks are emerging to make the IoT business model more compelling in 2019. IoT is the fundament of various organization i.e., Digital transformation that is capable to optimize the existing operation in order to create or pursue an innovative business models.

However, the potential features of IoT are still inadequate in literature to address the broad coverage of business models i.e., specifically for different industrial application domains (Gubbi et al., 2013). Because of IoT novelties and its related ambiguities, this inadequacy is chiefly instigated for the societal benefits. However, the corrective assessment and its possible solution may provide better warrant for IoT evolution that finds oneself to promote as an entrepreneur to create a value-added business model in order to captivate the commercial market. According to Osterwalder and Pigneur (2010), a business model is the key element of any industrial concern to realize the '*Creativity, Deliverability, and*

Customer Captivity' that describes the operational goals of the concern. In general, the business models use various key perspectives to analyze the business objectives. However, a well-known business model called as Canvas is basically preferred to inspect the interdisciplinary characteristic of two or more industrial concerns. This model is categorized into nine building blocks that deal with different elemental types to analyze its associated dependencies.

The canvas model is completed based on meta-analysis framework (Dijkman et al., 2015) where the other business model can also be coexisted with at least same number of building blocks moderately (Fan and Guang-Zhao, 2011; Liu and Wei, 2010) or entirely (Bucherer and Dieter, 2011; Sun et al., 2012) to analyze the significance of any application domain. As a result, this model strongly asserts that it could be subjected to discover the essential features using theoretical framework modeling. To substantiate the modeling process, the business IoT platform is literally focused that broadly studies the commercial objectives of IoT application domains to address the convergence features. A term known as IoT platform is preferred to express the compactness in terms of industrial contribution and customer satisfaction. Moreover, this platform is playing a crucial role as the middleware of IoT applications that creates bondage between the communication devices and the IT enterprises to fulfill the vision of IoT platform.

Multifarious challenges are exponentially growing on Today's manufacturing companies that include interdisciplinary approach, technological variance, and innovative strategies. To offer better product customization, the heterogeneous data related to the manufacturing process are gathered through the integration of smart manufacturing systems. Today's manufacturers connect the smart intelligence system to optimize the supply process (Papazoglou and Elgammal, 2017). This intelligence has attracted the vision of Industry 4.0 that focuses on the development of intelligent systems (Brettel et al., 2014; Kapetaniou et al., 2018). It is successively converged the technological operations and communication technologies to adjust the supply-chain management dynamically in order to achieve on-time customer-driven design and development. In the business aspect, Industry 4.0 has promoted the growth of manufacturing process that builds an end-to-end manufacturing system to coordinate with the real-time entities such as people, manufacturer, operation, sensor, and actuator (Davis et al., 2012; Zheng et al., 2018). In addition, the physical productivity should be collaborated with the supply-chain management to achieve better optimization in terms of product distribution (Papazoglou and Elgammal, 2017).

IoT and web of things (WoT) has gained the researches attention for smart manufacturing process that includes data collection, storage, elaboration, and action to analyze or share the generated values to-do supply-chain process. It is widely known as technology enabler (TE). However, without device automation, there is no other option to adjust the environmental values to participate in the productive process. As a result, interoperability based intelligent system plays a significant role to enable the TE that initiates the process of IoT to ease the need of human-machine interaction. Herman et al. introduced the design principles for industrie 4.0 (Hermann, Pentek, and Otto, 2016).

11.1.1 *IoT DEFINITION AND CHARACTERISTICS*

In this subsection, the primary sets of data parameters are embarked to deliberate the definitions of IoT that are discussed in three major aspects. Atzori et al. (2010) define the first aspect to state the IoT convergence in the perception of three dimensions such as *'Things Oriented,' 'Internet Oriented'* and *'Semantic-based.'* Initially, IoT was semantically introduced as *'World-Wide Networks of Inter Connected Objects'* to approach the IoT vantage point known as *'A Pervasive presence'* that provides the interaction among the network of objects to achieve the common objectives of IoT. ITU (2005, 2012) formulates the second aspect to propose that *'IoT is skilled to identity and integrate the physical objects into the communication networks.'* Lastly, the European Commission (Guillemin and Friess, 2009) conceptually defines that IoT has a dynamic-global network infrastructure to integrate the physical entities to experience the future networks, where the physical entities can have same identities, attributes, personalities, and intellectual interfaces. A repetitive term called *'things'* focuses a new facet for human interaction with application domain that enables the people and real-time object to connect each other in order to share the information through the knowledge of any routing service (Baldini et al., 2016).

The other related aspects represent some important characteristics to overlap on dynamic networks, global infrastructure, object identification, presence preservation, time spanning, storage data, network path, device interconnectivity, and interaction. The objective of IoT is to make an efficient real-time information system that shares the communication among the autonomous networks (Yang, Yang, and Plotnick, 2013). As a consequence, the emerging technologies are quantitatively reviewed that primarily focus on IoT application, security, and privacy issue, i.e., for pervasiveness of intelligent communication devices. Internet of devices connects with smart objects to sense and actuate the physical environment (Ng, 2014a). The purpose of Internet connectivity is to uniquely identify access and verify the device communication. A digital shadow or virtual-representation uses cyberspace to store the data that enables the interaction between the humans and the physical objects (Zhou and Piramuthu, 2015).

The physical objects may establish its communication to make the Internet service more ubiquitous and immersive that is completely based on object oriented design concepts. To envision the global network connection, machines, and devices are capable of interaction and interconnectivity. This technological advancement enhances the device communication over Internet access to achieve the desirable goals such as tracking, locating, identifying, and managing the physical objects. Moreover, the object interconnectivity converge the physical and the digital world to extend the constant benefits namely remote control access, data sharing and constant Internet access. Importantly, IoT emerges the global information services to extend the Internet infrastructure in order to realize the significance of information technologies, transmission, application, and acquisition of any communication systems. Above all, the context-aware (CA) objects i.e., physical, and virtual communicator provide a cooperative access to meet the standard demands of social and business requirements.

11.1.2   IoT: A PROMISING TECHNOLOGICAL PARADIGMS

IoT technologies prognosticate industrial evolution to connect the physical objects at any-time and anywhere through the use of routing/path establishment/service connection (Baldini et al., 2016; Guillemin and Friess, 2009; Man, Na, and Kit, 2015; UK Research Council, 2013). The IoT vision is of creating a smart world system that is composed of sensors and smart network components. The current IoT feature is named as Web3.0, which makes the users to interact with the physical environment more deeply than its antecedent Web 2.0 that heavily loads the physical environment with the technological convergence to go beyond content creation, discovery, and sharing (Kreps and Kimppa, 2015). Unsurprisingly, such a far-reaching vision of IoT has attracted the attention of both industrial practitioners and academic experts that underpin the innovative application and its related services. This significance is expected to provide a societal and business benefit that influences on individual and corporate policies to challenge new service ventures (Shin, 2014; Stankovic, 2014). On the other hand, it is evident that IoT would never be free from disputes and caveats. As an instance, IoT provides a pervasiveness and voluminous data generation to concern on privacy impingement in the device connectivity.

More research projects have been executed for the IoT application domains that encompass of eleven significant concepts to illustrate the significant features of information technologies and design infrastructures e.g., ubiquitous computing and semantic web. Lately, the IoT applications and its industrial concerns have been addressed for the state-of-the-art approaches and the future research guidelines (Olson, Nolin, and Nelhans, 2015). As to the best of my knowledge, very few research works have been published, however, none of them have analyzed the business perspective models. Atzori et al. (2010) published a review articles to present the vision and conceptual ideas of IoT enabling technologies, framework applications and potential future directions. Similarly, Li et al. (2014) provided an integration of IoT service-oriented architecture that enables technical standardization, challenges, innovations, development strategies, security, and privacy issues to address IoT based approaches towards possible solutions. Yan et al. (2015) showed the co-world analysis that utilizes the most prominent connectivity, i.e., wireless sensor networks (WSNs), radio frequency identification (RFID), etc. In addition, Mishra et al. (2016) applied the frequency analysis to address usage of sensor devices that reports RFID technology as more prominent rather than WSNs. Most of the research publications represent the above technologies in the form of clustering to overcome the security and privacy issues.

11.1.3   IoT: AN EXEMPLARY APPROACH

IoT standardizes the core idea of emerging technologies such as clothing, microwaves, and pacemaker that acquires the Internet access to sense and establish the communication between the physical objects (Gubbi et al., 2013). The potential use of IoT is to create a smart communication that senses the physical environment to interact with external systems. Essentially, IoT applies the business modeling techniques to develop various industrial applications such as agriculture, manufacturing, logistics, and healthcare (Dlodlo et al., 2012). Since the business modeling is capable to provide an interaction between the intelligent objects, the device connectivity or IoT interactivity is simply called an intelligent or smart system. The smart objects include connection establishment, device identification, and network configuration between the physical objects. A key enabler of IoT is simply called sensor technologies that detect various sensing parameters namely light, pressure, temperature, pressure, sound, and motion. The sensing parameters use Internet accessibility and controllability to provide the remote access that allows machine or device to update or fix the system errors remotely, whereby the time and money can substantially be saved (Figure 11.1).

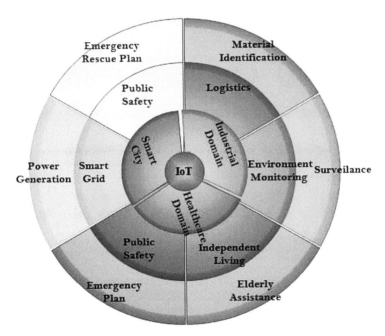

FIGURE 11.1 IoT application domains and its convergence areas.

Figure 11.1 illustrates the IoT Application Domains and Its Convergence Areas. Commonly, IoT has various application areas such as smart city, industrial, and healthcare domains that diversifies the research fields in the area of agriculture, logistics, manufacturing, smart grid, public safety, etc. In literature, IoT research has recommended five different cluster approaches. The first approach theoretically focuses on IoT challenges, whereas the second concerns on IoT implementation; the third applies the industrial logistics and supply chain management (SCM) systems; the fourth focuses on IoT design concepts; and finally, the fifth focuses on security and privacy issues related to IoT (Andersson and Lars-Gunnar, 2015). In order to fulfill the above approaches, the significant challenges namely interoperability and standardization are addressed. Basically, IoT elements are classified into hardware (i.e., sensor, network, and system hardware), middleware (i.e., data analysis and storage), and visualization tools. The IoT elements integrate the communication systems such as service-oriented architecture, peer-to-peer networking, cloud system, Wi-Fi, Bluetooth, WSNs and near field communication (NFC) technologies. Figure 11.2 shows the classification of IoT elements for interoperability and standardization.

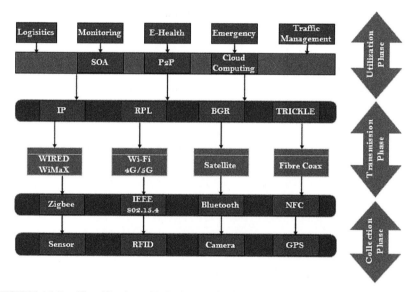

FIGURE 11.2   Classification of IoT element for interoperability and standardization.

11.1.4   BUSINESS MODEL DEVELOPMENT

Three main objectives such as create, delivers, and capture values are stated in the business model. It is specifically focused on *'the blueprint of company business aspects'* that describes: 1. What kind of service or product does the company offer?; 2. How does the company manufacture a product to deliver eventually?; and 3. How does the company get the license to sell the product?. This modeling may unite the technological advancement and economic-value creation to design nine key components that are as follows:

1. Who may be the customers?
2. What product is being sold?
3. How is being manufactured?
4. How the product is being sold?

Using Canvas business model, the analysis and creation tools are generally comprised of:

1. **Key Partners:** Who is partnering with the company?
2. **Key Activities:** What are the business activities required to deliver?
3. **Key Resources:** What are the resources necessitated to create?
4. **Value Proposition:** What values does the company want to fix?
5. **Customer Relationship:** What kind of relationship does the company have to create and maintain?
6. **Channels:** How does the company approach its customers?
7. **Customer Segment:** How does the company weigh the value?
8. **Cost Structure:** How does the company structure its business model?
9. **Revenue:** How does the income generate by the company?

The above components are referred as the business building blocks that create several elements to include the basic types such as convenient to access, low cost, more comfort, secure, etc. The business model creation allows sequential of communication processes that comprises of understands, design, implement, mobilize, and manage. In order to identify and gather the business resources, a successful modeling project should communicate rationally. As a consequence, the understand phase creates a business model to choose the design team; the design and implement phase constitutes the business objectives; and the mobilize and manage phase monitors the marketing condition to identify the customer objectives.

Osterwalder and Pigneur (2016) designed a business model to illustrate the change in business environment. Today's world addresses technological and political changes to explore the business development model that is highly demanded to change the market requirements. Bucherer and Uckelmann (2011) proposed a development of business model to bridge the gap between economic innovation and technological advancement. Pisano et al. (2015) created the business development model to improvise the business goals. As an instance, Osterwalder, and Pigneur (2016) discussed an iterative process for business development. Sosna et al. (2010) pointed out the trial and error approach to obtain the business goals. Teece (2010) posited the business model creation using learning approaches.

As a result, an ecosystem-based business model should be considered to evaluate the product iteration, business hypothesis, and company revenue. To analyze further, a methodology known as business, innovation, and customer experience (BIC) is considered. In addition, a technique of business model cliché is employed to distinguish the common pattern between the similar business models. It is explained about two business activities that are as follows: 1. How does the company create a business model to meet the customer requirement?; and 2. How does the company initiate the operational process to obtain the objectives. To achieve the above objectives, the business model cliché is categorized into three types, namely product, interaction, and resource. This categorization identifies the product cliché to meet the desirable goals of company and customer. The innovative epicenter builds the blocks of business model to create an innovative project that achieves three beneficial factors such as infrastructure, finance driven and customer. As an instance, a technology known as infrastructure-driven epicenter (IDE) includes the business blocks of Canvas such as key activities, partner, resource, and innovation, i.e., for 3D-prinitng (Pisano, Marco, and Alison, 2015).

The subjective perception and experience are referred to own the service or product that relates the emotional and experimental aspects to ease the customer use. As the user experience is more dynamic to change its business nature, it is very essential that the new customer experience changes the product or services to signify whether the customer will continue to use the same or change the product. Moreover, the BIC methodology uses to analyze the operation of corporate companies that recognizes the Canvas business model to mutually support the nine building blocks. The trend and similarities consider the framework objectives to achieve better business development process.

11.1.5   IoT BUSINESS MODELS

IoT has transformed various application areas for the wealth of business opportunities (Markman, 2015). This business model gives the IoT opportunity to under-represent the challenges of IoT. Mishra et al. (2016)

addressed the research expectation and creative models to highlight the dissimilarities of normal product and services. Borgia (2014) discussed the conventional business model that cannot be practiced to state the IoT context. Despite with the nonexistence of creative business model, it is ascertained that the new business model can be emerged when Internet service is surfaced. Mckinsey & Company posited to set the business model in order to provide a creative IoT. The examples of IoT business setting include smart vehicle, city, home, human interaction, and automation factory. A business-to-business application generates a consumer application that has raised up to 70% for the growth IoT convergence.

Since business market is facing the complication of IoT technological supplier, it is evident that it has a lack of standardization and competitiveness to achieve technology transfer, data storage and software platforms in order to provide a technological solution. In the growth of technology, software data in terms of data storage is expected to upsurge proportionally over time. In IoT market, the technologies such as computer and Internet are composed of three development phases: 1. The first phase designs a system infrastructure such as operating system or microprocessor; 2. The second phase integrates the application system such as online search; and 3. The third phase incorporates the system infrastructure and application to build a e-commerce business. The device performance is analyzed to track the product usage that engages in the trade of business models.

However, the modern business model cannot be more profitable than the conventional where the product owners transfer their product to the customers. The transfer ownership experiences on *'product-as-a-service (PaaS)'* to evaluate the product usage. Moreover, the product review provides user experience to-do maintenance and application updates that creates insightful thoughts to challenge the competitors. Generally, the business model provides product relationship to retain the company ownership that collects the user experience to track the system performance (Porter and James, 2015). In addition, data gathering and analysis enables the system experts to understand the significance of product development that ramifies the market dissection to derive the usage pattern. It is strongly assert that a customized product can be designed and developed for the customers. Over the development process, a significant change can be brought into the existence to create a new IoT opportunity i.e., for value creation and customer attraction.

Bucherer and Uckelmann (2011) illustrate various businesses IoT modeling to enable the significance of product enrichment. The first design uses PaaS to offer a powerful tool to show off the customer experience. The second design uses information service providers (ISPs) to provide the information related to finance, marketing, politics, business news and social trends. Importantly, IoT empowers the ISP benefits to collect the massive amount of data at low-cost. The third design connects the product life cycle model to provide the relationship between the customers and the co-assessment processes. IoT provides the product information to learn the customer co-activities that can be gathered either directly or indirectly. The fourth design closely connects with business analysis and decision making to enable the possibilities of data analysis and accessibility over supply-chain management system.

IoT analytics includes data management and security to provide a collaborative key partner that is pertained to product usage, application status, object location, and product version. Moreover, the above analytics may exchange directly or indirectly to analyze the views of information provider, companies, and end users. The information flow carries uni-directional or bi-directional or multi-directional to offer different channel interfaces such as business-to-business (B2B) and business-to-customer (B2C). In addition, the business transaction related to product information is characterized into product and money that provides business opportunities for third party retailers. IoT is still unstandardized to experience the interoperable issue that challenges to design IoT business models. However, the business opportunity is having an existence of object connectivity to satisfy the conditional need of the customers. In the end, the business models within IoT interconnect the possible development process to provide limitless service.

11.1.6   *IoT PLATFORM AND NETWORK PERSPECTIVE*

A business model based on IoT platform refers the design and development of application systems that form a customer or participant group to connect the device on one-end and IT enterprise on another-end. IoT platform has currently been in the existence of IT market for a decade; however the platform convergence has evolved in the IT enterprise before four years ago (IoT Analytics, 2015). The major role of IoT platform is to enable the object connectivity between the IoT devices with different application interfaces. The context of application is referred as visualization, user interface and diagnosis tool that controls and manages the device connectivity (Kaa Project, 2017). As a result, the IoT platform strongly provides a communication interface between the connective device, application developer, and IT enterprise. Specifically, a middleware system is enabled to provide a constructive interface between application and technology (Balamuralidhar, Prateep, and Arpan, 2013). This evolution has undergone for traditional machine-to-machine (M2M) communication that is integrated to design a collective application system containing a variety of application developers, device connection and IT enterprise systems (Figure 11.3).

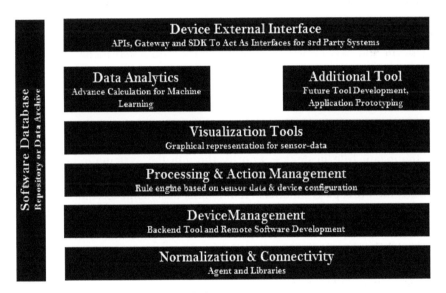

FIGURE 11.3 Components of IoT platform.

However, this convergence platform is so essential to create a scalable application system that connects things, systems, and individual. As a consequence, this communication infrastructure is considered to improve the device interactivities, which is assessed to reach 1 billion dollars by 2019 (Scully, 2016b). Specifically, IoT platform comprises of nine different functions that are as follows: (1) The first function refers to connectivity and normalization, which can be accomplished through the IoT application interfaces; (2) The second one deals with device management system, which is intended to remote server access and control; (3) The third refers to the database management to store the structured and unstructured data; (4) The fourth has processing management that utilizes the data collection to engage in decisive action; (5) The fifth provides data analytics through the use of data mining, ML and pattern detection; (6) The sixth is in charge of data visualization to analyze and process the raw data; (7) The seventh provides additional components to explore the IoT supplementary tools; (8) The eight one offers the external interfaces to enable the platform access through the software development kits (SDK); and (9) The ninth provides a customizable IoT platform to provide comprehensive implementation as illustrated in Figure 11.3.

Most of the existing IoT platform do not offer data security, cloud storage and connectivity management that are completely differed from each other in terms of segment focus, technological advancement and customization approach. This connectivity enables normalization, action platform and data collection to execute the rule-based management that accommodates numerous device connectivity to provide seamless integration. IoT platform supplies device integrity to offer *'one-stop-shop'* platform that has limited developer's tool and technical support. The components of IoT platform give a desirable characteristic of identity management to hold the feature of multi-tenancy. In addition, IoT plays a significant role in device security through the systematic approaches such as authentication, integrity, and access control. The core ecosystem of business model is called as technological convergence that comprises of building blocks, auxiliary product, and services to establish the device connectivity between the physical objects and the Internet access.

11.1.7   INDUSTRY 4.0

Of late, the academicians and the practitioners have given various key concepts for IoT domain for unstructured knowledge representation (Liu et al., 2018; Hussain and Cambria, 2018). Lelli et al. (2019) represented the abstraction model that shows the generic instrument and its related elemental solutions. The connective challenges of scientific instruments with computation infrastructure show the interoperability issues controlling or monitoring using IoT devices. They conceptualize the device interactivities more operational and abstract to devise a set of communication parameters, execution commands, control model, and web-based description that could be incorporated with semantic digital transformation. A solution known as instrument element deals with heterogeneous and distributed computing services to offer an insight in the construction of structure blocks i.e., for complex IoT systems as shown in Figure 11.4.

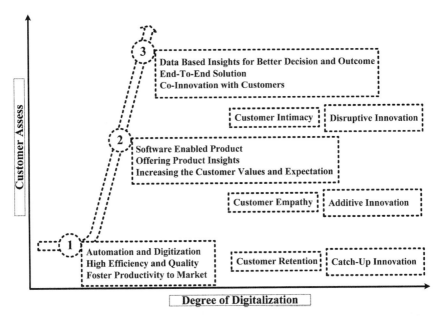

FIGURE 11.4   Instrument elements dealing with heterogeneous and distributed computing services.

The emerging technologies are intended to provide prototype models and proof of concepts that focuses on the development of semantic web-stack. In literature (Patel, Ali, and Sheth, 2018; Andročec, Novak, and Oreški, 2018), Web Stack-IoT is not widely employed to achieve intelligent interoperability. As a result, it is evident that it could not envision the scope of Industry 4.0. However, most of the researches prefer to use semantic stack to develop a case-study model (Cheng et al., 2018; Lin and Yang, 2018). The differential conflict is focalized on the development of web-stack application to generalize the key concepts of IoT infrastructure. It is to note that the field of web-stack application is emerging towards the evolution of web-stack of things (WSoT) that intend to describe the significance of current trends without structured categorization. Datta et al. (2015) outlined the relevance of IoT development to adopt the feature of intelligent interoperability, i.e., for industry 4.0. It is also noted that the evolution of IoT fosters the development of IoT platform to assess the level of device connectivity. In addition, a set of criteria is defined to assess the web-stack interoperability using IoT model. Patel et al. (2018) demonstrated a domain specific IoT model to discuss the importance of artificial intelligence (AI) in Industry 4.0. Table 11.1 shows the comparative analysis of IoT design model for the development of web-stack in industry 4.0. From Table 11.1, it is observed that the intelligent interoperability is highlighted more to achieve the industry 4.0 design model.

TABLE 11.1   Comparative Analysis of IoT Design Model for Development of Web-Stack in Industry 4.0

Author	Description	Key Findings
Noy et al. (2001)	Illustrates the stepwise approach for web-stack IoT design	Development of web-stack methodology
Compton et al. (2009)	Reviews the importance of sensor technologies	Web-stack properties and its desirable capabilities
Figueroa (2010)	Focuses on the building blocks of web-stack networks	Web-stack creation
Hachem (2011)	Presents the challenge of IoT model including scalability, heterogeneity, and unknown network-topology	Solving interoperability issues
Gyrard (2015)	Presents best practice design for semantic web-stack community	Web-stack design, Recommendation tools and metrics evaluation
Ganzha (2017)	Illustrates the design concepts of web-stack processing model	Provides IoT use case models i.e., for smart healthcare and transportation
Androcec (2018)	Addresses the interoperability issues in the domain of IoT business model	Importance of IoT modeling domain
Lelli (2019)	Enables dynamic industry 4.0	Intelligent interoperability to enable dynamic reconfiguration

11.2   METHODOLOGIES

The goal of this chapter is to create a framework model that develops an intellectual business models for IoT applications to build a building blocks in order to specify the blocking types. Therefore, an empirical research study is preferred to identify the building block that chooses an exploratory research design to-do data qualitative approach. To make the concept more holistic behavior for IoT, this chapter discusses the complexity of socio-technological systems. This perception comprises of physical, technological, and socio-economic environment. The physical environment involves human and machine interaction with the help of ubiquitous network to enable auto-communication and interaction. The technological environment includes hardware, software, network, data, and technical standards. The socio-technological system composes of various stakeholder activities including entrepreneurs, and business leader who are constantly engaged in entre-intrapreneurship development process within the IoT platform design. Figure 11.5 depicts the physical environment for IoT application systems.

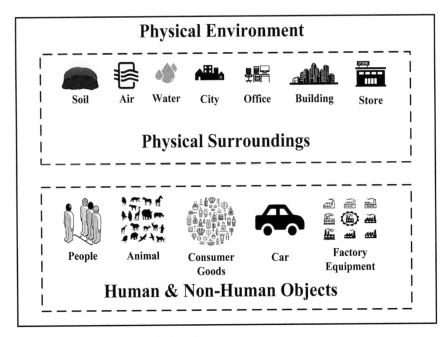

FIGURE 11.5   IoT as a physical environment.

11.2.1   *PHYSICAL ENVIRONMENT*

In physical environment, the machines are configured with the help of wireless communication devices to automate the machinery process. The physical objects are the communication users to provide a direct interaction using various accessing systems such as smartphones, laptops, medical sensors, RFID-tag, etc. Moreover, the objects such as car, train, flight, etc. integrate with the smart device to establish the connection over wireless device to send or share the data to the distributed server. The physical and non-physical objects are integrated to build a friendly ecosystem that comprises of physical surroundings such as house, office building, park, city, etc.

The surrounding can also be assumed as soil, sand, and air. In any case, the above environmental system can connect the physical objects to the IoT platform that embeds several sensors to measure the ecological properties namely temperature, humidity, pressure, soil acidity etc. Alternatively, a physical space can be considered to sense the elements namely building, doors, and signs that interact with wireless devices to analyze the ecological characteristics. As an instance, an office premise is equipped with RFID reader to sense the physical object integrated with badge transmitter that makes the door to open automatically when any authentic physical object enters into the vicinity of transmission region.

11.2.2   TECHNOLOGICAL ENVIRONMENT

The technological environment comprises of hardware, software, network, data, and technical standards to enable the physical object interaction. The hardware integrates with the IoT objects to connect the portable devices e.g., smartphone, and tablet that interacts with the wearable devices e.g., Apple watch. It is used to transmit and receive the information wirelessly. Wearable sensors or RFID transponder interface a communication tag that can be deployed for different types of application domains. For example, RFID uses active or passive power tag to enable more advanced features such as GPS functionalities. Moreover, this communication tag generates a unique identifier to retrieve the data from the centralized database. Some communication tag senses, transmits, and stores the data from the internal data storage that communicate with intelligent hardware to access over Internet. As an instance, a refrigerator equips with network interface card that allows the physical object to communicate over Internet between the IoT nodes. Figure 11.6 illustrates the technological environment of IoT application.

In general, IoT software is classified into application software and middleware. The former considers client-side application and server-side software to support the consumer requirements. For example, an Uber's

App gives the opportunity to smartphone to gain the services e.g., trip-request, location-sharing, and driver contact. This configuration setup allows the user to establish the communication and navigation service, whereby users can transit from one place to another to explore the features such as payment process and trip-schedule. However, some IoT application may rely on middleware to ease the need of software component in order to establish the communication between the IoT devices. Usually, different kind of networks offer object connectivity i.e., between the human and the non-human objects to allow the smart device to connect and share data located within the proximity. On the other hand, the physical objects may establish its network connectivity over Wi-Fi network to provide campus or corporate area networks.

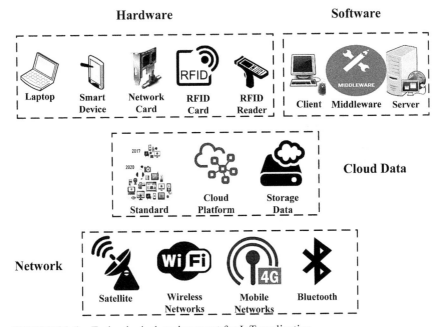

FIGURE 11.6 Technological environment for IoT application.

Internet-based IoT application integrates cloud to provide a seamless hardware platform that allows data storage and analytics for different kind of embedded application systems. As an instance, Microsoft Azure offers a global data center networks to provide Internet access i.e., for business IoT platform. IoT exponentially enhances the storage of data that is simply

existence versus customer experience and comments from individual versus human interaction. Figure 11.8 shows differential behavior of IoT devices and services. This approach is systematically approached to meet the customer requirements that provide better understandability to improve the efficiency of business. To broadly view the IoT market segmentation of industry, the different service sectors are: 1. Energy; 2. Building; 3. Home; 4. Consumer; 5. Life science and healthcare; 6. Transportation; 7. Retail; 8. IT & Networks; and 9. Industries.

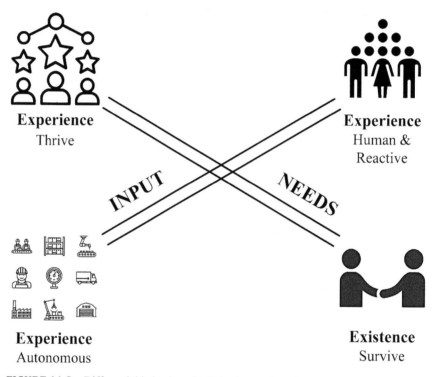

FIGURE 11.8   Differential behavior of IoT devices and services.

11.3.2   *CUSTOMER RELATIONSHIPS*

Most of the IT companies search for the rightful business model to experience that includes state-of-the-art approaches to incorporate new ideas and futuristic concepts. The customer relationship is deeply interviewed to analyze the frequent communication between the corporate employees and the customers. The company examines the interview questions to design the marketing strategies that provide an optimal path to consult with the customers. 'Even though there are a lot of academic institutes to educate the people, there still exist of IT challenges to include how does the institute makes a money using IoT, how do you explores the IT strategies and what is the business policy preferred to improve the educational system.

The customer expects to implement the business IoT that introduces the fundamental changes in business revolution. In business IoT platform, the original equipment manufacturers (OEMs) deeply relate the business activities to build the blocks of customer relationship. According to the business building blocks, self-service, and service automation are particularly selected to draw the consultative relationship between customer and company. Dijkman (2015) suggests various findings to provide the customer relationship, however, this chapter addresses the consultative relationship to address the significance of business building blocks.

11.3.3 REVENUE STREAMS

In IoT platform, various revenue streams are chosen to connect the business modeling that concerns on financial viability to define the cost structure. To secure the business profitability over time, an appropriate risk management is considered that provides a financial gain to exercise high delivery cost. It is apparent that the value-added dimension is provisionally enabled to discuss on cost and revenue. In digital business model, an improved cost efficiency can be realized to streamline the delivery process i.e., for the development of digital business model. In fact, the use of digital technology can improve the price margin to provide the business opportunities such as pay-per-use, fixed price, creating the product values, etc.

To implement the business model, an interdependency ecosystem is represented that increases the uncertainties to handle the risk-management approaches. However, the key benefits are to create the digital transparency in the process of customer relationship to provide the insightful thoughts. Moreover, the outcome based services offer business opportunities to closely connect the digital transparency that achieve the desirable goals of IoT business models.

11.3.4   *KEY RESOURCES, ACTIVITIES, AND PARTNERS*

A set of key resources is opted to make up the building blocks of business industries that include the IoT resources such as Exosite, Ubidots, and a data center of Cumulocity. Dijkman (2015) shows the software capabilities to include the types of business relation, financial resource, and intellectual property. To deliver the business IoT model, different kinds of key resources are chosen that enables: 1. Physical Resource; 2. Human Resource; 3. Financial Resource; and 4. Intellectual Resource. These are essential to achieve the key activities of business model that are considered to motivate the creative partnership. They are: 1. Risk reduction and uncertainty; 2. Scaling economy and business optimization; and 3. Resource acquisition and organization activities. Figure 11.9 illustrates the business canvas model.

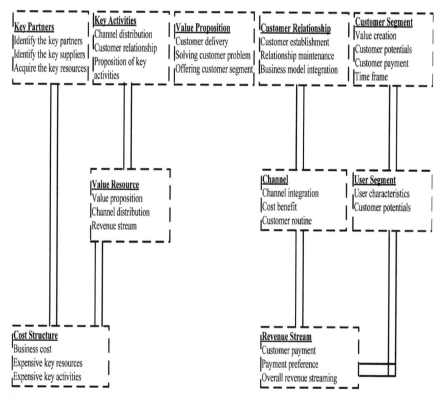

FIGURE 11.9   Business canvas model.

11.3.5 PRESENTATION FINDINGS

This study is aimed to provide a sequential exploratory study that defines the relationship when the key variables are unknown. In this study, qualitative data are gathered and analyzed to integrate two different approaches such as constructive validation and strong conclusion. Importantly, the suitable building blocks are identified to realistically work on business IoT model i.e., through the knowledge of literature and professional interviews. In addition, a relative strategy is used to determine the suitable building blocks that discuss the challenges of business relevance.

In line with the above motivational objectives, an IoT business model known as Canvas is chosen as starting point to develop an interview centric framework protocol. As the interview framework is based on semi-structured, the practitioner viewpoints are aimed to work on the correctness of building blocks and its related business model provided in Canvas. The survey questions are literally related to the questionnaire discussed in (Osterwalder and Pigneur, 2010). In practice, the participants are examined through some specific viewpoints that are as follows:

1. A recommendation obtained from the business network;
2. A recommendation obtained from the IoT experts;
3. An IoT company reference observed on Internet;
4. A recommendation obtained from the previous interview candidates.

Table 11.2 shows the descriptive statistics for each company's interview that shows various industrial sectors to classify whether the company size, operation process, and client type offer the IoT business strategy to consider the interview activities as shown in Table 11.3. It is primarily focused on product, service or both to identify the year of experience in the market. In general, the interviews are recorded to verify the activities of interviewee that adapts different types of IoT business model to-do addition, deletion, division, or unification in order to explore an alternative classification approach. In addition, it shows how often particular activities recur during the interview process.

As referred in (Dillman, 2000), a Likert scale is preferred to verify the opinion of IoT business professions that divulges the importance of building blocks and its related types. The general opinion polls are the vote of respondent that are 'Strongly Agree (A),' 'Agree (B),' 'Strongly Disagree (C)' and 'Disagree (D)' to show off their preference options on the IoT building blocks. Therefore, it could eventually be incorporated

TABLE 11.2 Descriptive Statistics for Each Company's Interview

Sectors	Number of Industries
Agriculture	1
Energy conservation	1
Healthcare or Hospital	2
Smart home	2
Smart building or construction	1
Supply chain	3
Intelligent Transportation	1
Size	**Number of Industries**
Micro i.e.,< 10 *employee*	5
Small, i.e., 10 *to* 50 *employee*	2
Medium, i.e., 50 *to* 250 *employee*	2
Large, i.e., > 250 *employee*	2
Client	**Number of Industries**
Business-To-Business	8
Business-To-Consumer	3
Product or Service	**Number of Industries**
Product	4
Service	2
Both	5
Year of Experience	**Number of Industries**
< 1 Year	3
1 – 5 Years	6
5 – 10 Years	1
> 10 Years	1

into business IoT model. To understand the various potential problems, a rigorous analysis known as *'IP Double Checking'* is considered that can reveal the intentions of the respondent i.e., one-time opinion or more. In order to make this practice more reliable, a timeframe is scheduled i.e., ~15 mins. It means that if any respondent opinion falls below the timeframe, then his/her view will not be considered to process in this approach. Specifically, a standard deviation is calculated for each building block that shows its significance in term of influential factors. Assume that the influential factor is ⟨0⟩, it implies that the respondent fills all the details promptly. While surviving the above viewpoints, 199 responses were collected. Of which 96 responses were completed intact with the questionnaire and constraints whereas 103 responses were reported partially because of incompleteness, suspiciousness, and subversion of timeframe.

TABLE 11.3   Interview Activities by Industry Sector

Number of Interviews	Name of the Company	Sectors	Application of IoT	Findings
1	Focus-Cura	Healthcare	"'ThuisMeetApp'"	A
2	Dutch-Domotics	Healthcare	"'Zorgdomotica'"	A
3	Hoogendoorn	Agriculture	"iSii compact"	C
4	Essent	Energy	"e-Thermostat"	A
5	Bundles	Smart home	"Washbundles"	D
6	Blinq-system	Smart building or construction	"MapIQ"	A
7	Ambient-system	Supply chain	"Ambient supply chain"	B
8	GSETrack	Intelligent Transportation	"GSETrack"	C
9	Prometheus	Supply chain	"Telematica"	B
10	Philips	Smart home	"Philips Hue"	A
11	Alexander & Mieloo	Supply chain	"ScanGreen"	A

Subsequently, an exploratory study was done based on the above analysis to examine the ignored factors padding under the building blocks or not with the modified building block types such as key partners, activities, resources, and cost structure. In all the cases, a measurement technique known as 'Kaiser-Meyer-Olkin' is used to verify the adequacy rate of sampling, i.e., range from $\langle 0.55 \rangle$ to $\langle 0.70 \rangle$ (Papazoglou and Elgammal, 2017). Moreover, to analyze the process further, a post-estimation is preferred between the above ranges of factors. Thus, a promax or oblique is applied to ease the factor interpretation process instead of orthogonal-rotation. Figure 11.10 shows that a sequential exploratory study achieves 94.5% confidence interval as the average score overall IoT building blocks. Also, the opinion poll survey points out the significance of IoT building blocks to analyze its value proposition factor that is recorded to be higher score $\bar{x} \approx 6.38$ in comparison with all other building blocks. Moreover, the measurement difference between the building blocks is low, though the channels significance levels are considered to be very less ~ < 0.02.

Notably, the interviewees building blocks indicate the value proposition as ⟨8 *Times*⟩, customer-relationship as ⟨5 *Times*⟩ and key partners as ⟨4 *Times*⟩ for the building blocks of IoT model. Thus, a sequential exploratory study reports that the interviews coexist with the survey.

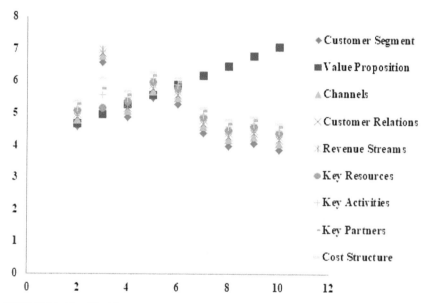

FIGURE 11.10   Significance of building blocks using sequential exploratory study.

11.4   DISCUSSIONS

From the opinion polling, it is understood that the value proposition achieves significant results for the building blocks of business IoT model. It is also found that the interviewees have given more preference for value proposition rather other building blocks. In addition, the parameters such as value proposition, key partnership, and customer relationship are carefully considered to modulate the execution of three building blocks to determine the specific nature of business model. In value proposition block, the respondent sensibly judges the user convenience, model performance, comfortability, job completeness, and chance of updates to explore the revenue model. In key partnership, the evidence of hardware partner, software developer, customer launch and data pattern analysis is cautiously chosen to give the product portfolio in order to acquire the business outsourcing. In customer relationship, the discontinuation of building block is noticed to focus on community formation and co-existence. This inference provides an effective building blocks, personalization, and information context realizing over the period of time. Moreover, the effective communication enables more customer contacts to improve the self-reliance.

11.4.1 PRODUCT AS A BUSINESS MODEL

A business model is called as a framework tool, which provides a descriptive value of a company to offer one or more customer segments. This firm approach develops a systematic architecture that creates, markets, and delivers a business product to generate profitable revenue to sustain in the competitive market (Osterwalder and Pigneur, 2010; Dillman, 2000). It is usually divided the building blocks into various components such as customer-segment, value-proposition, channels, customer-relationship, revenue-stream, key-resource, key-partnership, key-resource, cost-structure, and key-activities (Osterwalder and Pigneur, 2010). A framework tool provides a key assistance to develop a suitable business model that could be searched through the existing model to conduct a rigorous analysis. In this research study, the keywords such as business model and IoT were chosen to-do an extensive analysis on 'ACM Digital Library,' 'SpringerOpen,' 'ScienceDirect,' 'IOS Press,' "MDPI' and 'IEEE.' From the search analysis, it is asserted that IoT can enrich the business to sell the product more reliably to meet the demand of the customers. In order to strengthen the assertion, the business model utilizes the usage-based revenue stream to obtain whether it is obtaining the higher margin or not. However, the customers are experiencing through the implementation of PaaS.

11.4.2 IoT CHALLENGES

This subsection highlights the benefits of challenges and recommendation for business IoT model. A habitual or challenging attitude is always appreciated to realize the choice of importance:

- Predict the nature of real time environment and its collaborative impact.
- Create a customer specific product to enable the product mainte-nance through data convergence.
- Provide revenue-stream to realize the use of product through customer experience.
- Understand how does the other business collaborate to form a new venture.
- Identify the customer expectation to design an optimistic business model.
- Realize the significance of target outcome and deployment.
- Investigate the current opportunities to leverage the use of data management.
- Make sure that the engineers involving in the process of develop-ment showing enough knowledge and skills.
- Embrace the feature of maintainability to manage the new business opportunities.

11.5   CONCLUSION

This chapter presents the exploratory study on the topic of business IoT platform that is subjected to conduct eleven semi-structured questions. This enablement platform centrally leads to understand the building blocks and types of business model to discover the importance of generic IoT applications. The insightful thought is provided to signify the difference between business and generic IoT application platform. The explorative and quantitative-based approach is considered to prevent the small-factor analysis. Moreover, a subjective concept of business IoT platform has been created using a sequential exploratory approach that provides a sequence of activities to build an intuitive business model. Using a rigorous analysis, an exploratory study was carried out to examine the details of 199 respon-dents that show the importance of building blocks in the business model.

Nevertheless, it is still unclear about the future outlook of business IoT model to explore intellectual opportunities. In order to understand better, a creative business IoT model was developed to specify the key elements of building blocks such as customer-segment, value-proposition, channels, customer-relationship, revenue-stream, key-resource, key-partnership, key-resource, cost-structure, and key-activities. To establish the existence of business framework strategy, a canvas model was proposed. This model was applied to determine *'how does the business IoT model configure?' and 'what does a company enforce to achieve the specific proposition of IoT applications.* 'The quantitative and qualitative approaches were identified to signify the importance of building blocks for IoT product design and development. Moreover, the business canvas model has been extended for the convergence of IoT product design and development. In future, there may be many possible creative methods and key factors recommended to identify the different revenue-streams to open up the challenging issues in the business IoT platform.

KEYWORDS

- **internet of things**
- **near field communication**
- **radio frequency identification**
- **technology enabler**
- **web of things**
- **wireless sensor networks**

REFERENCES

Anderson, J. C., Narus, J. A., & Van Rossum, W., (2006). Customer value propositions in business markets. *Harvard Business Review, 84*(3), pp. 91–99.

Andersson, P., & Lars-Gunnar, M., (2015). "Service innovations enabled by the "internet of things." *IMP Journal, 9*(1), 85–106.

Andročec, D., Novak, M., & Oreški, D., (2018). Using semantic web for internet of things interoperability: A systematic review. *International Journal on Semantic Web and Information Systems (IJSWIS), 14*(4), 147–171.

Atzori, L., Iera, A., & Morabito, G., (2010). The internet of things: A survey. *Comput. Netw. 54*, 2787–2805.

Balamuralidhar, P., Prateep, M., & Arpan, P., (2013). "Software platforms for internet of things and M2M." *Journal of the Indian Institute of Science, 93*(3), 487–498.

Baldini, G., Botterman, M., Neisse, R., & Tallacchini, M., (2016). Ethical design in the internet of things. *Sci. Eng. Ethics,* 1–21.

Borgia, E., (2014). "The internet of things vision: Key features, applications and open issues." *Computer Communications 54*, 1–31.

Brettel, M., Friederichsen, N., Keller, M., & Rosenberg, M., (2014). How virtualization, decentralization and network building change the manufacturing landscape: An industry 4.0 perspective. *International Journal of Mechanical, Industrial Science and Engineering, 8*(1), 37–44.

Bucherer, E., & Dieter, U., (2011). *"Business Models for the Internet of Things* (pp. 253–277)."* Architecting the internet of things. Springer Berlin Heidelberg.

Cheng, Y., Zhao, S., Cheng, B., Hou, S., Shi, Y., & Chen, J., (2018). Modeling and optimization for collaborative business process towards IoT applications. *Mobile Information Systems.*

CISCO and General Electric. *"Popular Internet of Things Forecast of 50 Billion Devices by 2020."* Archived @ https://spectrum.ieee.org/tech-talk/telecom/internet/popular-internet-of-things-forecast-of-50-billion-devices-by-2020-is-outdated (accessed on 16 February 2020).

Compton, M., Henson, C. A., Lefort, L., Neuhaus, H., & Sheth, A. P., (2009). *A Survey of the Semantic Specification of Sensors.*

Datta, S. K., Bonnet, C., Gyrard, A., Da Costa, R. P. F., & Boudaoud, K. (2015, October). "Applying Internet of Things for personalized healthcare in smart homes," In *2015 24th Wireless and Optical Communication Conference (WOCC)* (pp. 164–169). IEEE.

Davis, J., Edgar, T., Porter, J., Bernaden, J., & Sarli, M., (2012). Smart manufacturing, manufacturing intelligence and demand-dynamic performance. *Computers and Chemical Engineering, 47,* 145–156.

Dijkman, R. M., et al., (2015). "Business models for the internet of things." *International Journal of Information Management, 35*(6), 672–678.

Dillman, D. A., (2000). *Mail and Internet Surveys: The Tailored Design Method* (Vol. 2). New York, NY, USA: Wiley.

Dlodlo, N., et al., (2012). "The state of affairs in internet of things research." *Electronic Journal of Information Systems Evaluation, 15*(3), 244–259.

Fan, P., & Guang-Zhao, Z., (2011). "Analysis of the business model innovation of the technology of internet of things in postal logistics." *Industrial Engineering and Engineering Management (IE&EM).* IEEE 18Th International Conference. IEEE.

Ganzha, M., Paprzycki, M., Pawłowski, W., Szmeja, P., & Wasielewska, K., (2017). Semantic interoperability in the internet of things: An overview from the INTER-IoT perspective. *Journal of Network and Computer Applications, 81,* 111–124.

GS1, (2015). *The EPC Global Architecture Framework.* Available at: https://www.gs1.org/epcrfid-epcis-id-keys/epc-rfid-architecture-framework/1–7 (accessed on 16 February 2020).

Gubbi, J., et al., (2013). "Internet of things (IoT): A vision, architectural elements, and future directions." *Future Generation Computer Systems, 29*(7), 1645–1660.

Guillemin, P., & Friess, P., (2009). *Internet of Things Strategic Research Roadmap.* The cluster of European research projects. In Technical Report.

Gyrard, A., Serrano, M., & Atemezing, G. A., (2015). Semantic web methodologies, best practices, and ontology engineering applied to Internet of Things. In: *2015 IEEE 2nd World Forum on Internet of Things (WF-IoT)* (pp. 412–417). IEEE.

Hachem, S., Teixeira, T., & Issarny, V., (2011). Ontologies for the internet of things. In: *Proceedings of the 8th Middleware Doctoral Symposium* (p. 3). ACM.

Hermann, M., Pentek, T., & Otto, B., (2016). Design principles for Industrie 4.0 scenarios. In: *2016 49th Hawaii International Conference on System Sciences (HICSS)* (pp. 3928–3937). IEEE.

Hussain, A., & Cambria, E., (2018). Semi-supervised learning for big social data analysis. *Neurocomputing, 275,* 1662–1673.

IoT Analytics, (2015). *"IoT Platforms—The Central Backbone for the Internet of Things."* November. White Paper.

ITU Strategy and Policy Unit, (2005). *ITU Internet Reports 2005: The Internet of Things.* International Telecommunication Union (ITU), Geneva.

ITU-T, Y. 2060, (2012). Next generation networks-frameworks and functional architecture models. In: Series, Y., (ed.), *Global Information Infrastructure, Internet Protocol Aspects and Next-Generation Networks.* Telecommunication Standardization Sector of ITU.

Kaa Project, (2017). *"What is an IoT Platform?"* The Kaa Project.

Kapetaniou, C., Rieple, A., Pilkington, A., Frandsen, T., & Pisano, P., (2018). Building the layers of a new manufacturing taxonomy: How 3D printing is creating a new landscape of production eco-systems and competitive dynamics. *Technological Forecasting and Social Change, 128,* 22–35.

Kreps, D., & Kimppa, K., (2015). Theorizing web 3.0: ICTs in a changing society. *Inf. Technol. People, 28,* 726–741.

Lanza, J., Sanchez, L., Santana, J. R., Agarwal, R., Kefalakis, N., Grace, P., & Cirillo, F., (2018). Experimentation as a service over semantically interoperable internet of things test beds. *IEEE Access, 6,* 51607–51625.

Lelli, F., (2019). Interoperability of the time of industry 4.0 and the internet of things. *Future Internet, 11*(2), 36.

Li, S., Xu, L. D., & Zhao, S., (2014). The internet of things: A survey. *Inf. Syst. Front., 17,* 243–259.

Li, S., Xu, L. D., & Zhao, S., (2015). The internet of things: A survey. *Information Systems Frontiers, 17*(2), 243—259.

Lin, C. C., & Yang, J. W., (2018). Cost-efficient deployment of fog computing systems at logistics centers in industry 4.0. *IEEE Transactions on Industrial Informatics, 14*(10), 4603–4611.

Liu, L., & Wei, J., (2010). "Business model for drug supply chain based on the internet of things." *2010 2ⁿᵈ IEEE International Conference on Network Infrastructure and Digital Content.*

Liu, Y., Du, F., Sun, J., Jiang, Y., He, J., Zhu, T., & Sun, C., (2018). A Crowdsourcing-based topic model for service matchmaking in internet of things. *Future Generation Computer Systems, 87,* 186–197.

Man, L. C. K., Na, C. M., & Kit, N. C., (2015). IoT-based asset management system for healthcare-related industries. *Int. J. Eng. Bus. Manag., 7.*

Markman, J., (2015). "$19 trillion internet shockwave." *Money and Markets Financial Advice Financial Investment Newsletter.*

Mckinsey & Company. *"The Internet of Things: Mapping the Value Beyond the Hype."* Archived@https://www.mckinsey.com/~/media/mckinsey/business%20functions/ mckinsey%20digital/our%20insights/the%20internet%20of%20things%20the%20 value%20of%20digitizing%20the%20physical%20world/the-internet-of-things-mapping-the-value-beyond-the-hype.ashx (accessed on 16 February 2020).

Mishra, D., et al., (2016). "Vision, applications and future challenges of internet of things: A bibliometric study of the recent literature." *Industrial Management and Data Systems 116*(7), 1331–1355.

Ng, V., (2014a). The future of enterprise mobility-enterprise IoT? *Network World Asia, 11,* pp. 10–12.

Noy, N. F., & McGuinness, D. L., (2001). *Ontology Development 101: A Guide to Creating Your First Ontology.*

Olson, N., Nolin, J. M., & Nelhans, G., (2015). Semantic web, ubiquitous computing, or internet of things? A macro-analysis of scholarly publications. *J. Doc., 71*, 884–916.

Osterwalder, A., & Pigneur, Y., (2010). *Business Model Generation: A Handbook for Visionaries, Game Changers, and Challengers.* Hoboken, NJ: John Wiley & Sons.

Osterwalder, A., & Yves, P., (2010). *Business Model Generation: A Handbook for Visionaries, Game Changers, and Challengers.* John Wiley & Sons.

Osterwalder, A., (2016). "Competition is not part of your business model." *Business Model Alchemist.*

Papazoglou, M. P., & Elgammal, A., (2017). The manufacturing blueprint environment: Bringing intelligence into manufacturing. In: *2017 International Conference on Engineering, Technology and Innovation (ICE/ITMC)* (pp. 750–759). IEEE.

Patel, P., Ali, M. I., & Sheth, A., (2018). From raw data to smart manufacturing: AI and semantic web of things for industry 4.0. *IEEE Intelligent Systems, 33*(4), 79–86.

Pisano, P., Marco, P., & Alison, R., (2015). "Identify innovative business models: Can innovative business models enable players to react to ongoing or unpredictable trends?" *Entrepreneurship Research Journal, 5*(3), 181–199.

Porter, M. E., & James, E. H., (2015). "How smart, connected products are transforming companies." *Harvard Business Review, 93*(10), 96–114.

Scully, P., (2016b). "*5 Things to Know About the IoT Platform Ecosystem.*" IoT analytics-market insights for the internet of things.

Shin, D., (2014). A socio-technical framework for internet-of-things design: A human-centered design for the internet of things. *Telematics Inform., 31*, 519–531.

Sosna, M., Rosa, N., Trevinyo-Rodríguez, & Ramakrishna, S. V., (2010). "Business model innovation through trial-and-error learning: The Naturhouse case." *Long Range Planning, 43*(2), 383–407.

Stankovic, J. A., (2014). Research directions for the internet of things. *IEEE Internet Things J., 1*, 3–9.

Suárez-Figueroa, M. C., (2010). *NeOn Methodology for Building Ontology Networks: Specification, Scheduling and Reuse.* Doctoral dissertation, Informatica.

Sun, Y., et al., (2012). "A holistic approach to visualizing business models for the internet of things." *Communications in Mobile Computing, 1*(1), 1.

Teece, D. J., (2010). "Business models, business strategy and innovation." *Long Range Planning, 43*(2), 172–194.

UK Research Council, (2013). *Research in the Wild-Internet of Things 2013.*

Yan, B. N., Lee, T. S., & Lee, T. P., (2015). Mapping the intellectual structure of the internet of things (IoT) field (2000–2014): A co-word analysis. *Scientometrics, 105*, 1285–1300.

Yang, L., Yang, S. H., & Plotnick, L., (2013). How the internet of things technology enhances emergency response operations. *Technol. Forecast. Soc. Chang. 80*, 1854–1867.

Zheng, P., Wang, H., Sang, Z., Zhong, R. Y., Liu, Y., Liu, C., & Xu, X., (2018). Smart manufacturing systems for Industry 4.0: Conceptual framework, scenarios, and future perspectives. *Frontiers of Mechanical Engineering.* Higher Education Press. https://doi.org/10.1007/s11465–018–0499–5 (accessed on 16 February 2020).

Zhou, W., & Piramuthu, S., (2015). Information relevance model of customized privacy for IoT. *J. Bus. Ethics, 131*, 19–30.

CHAPTER 12

Social Media Analytics Using R Programming

SONIA SAINI,[1] RUCHIKA BATHLA,[1] and RITU PUNHANI[2]

[1]*Amity Institute of Information Technology, Amity University, Noida, Uttar Pradesh, India, E-mails: sonia.22.saini@gmail.com (S. Saini), bathla.ruchika@gmail.com (R. Bathla)*

[2]*Department of Information Technology, ASET, Amity University, Noida, Uttar Pradesh, India, E-mail: ritupunhani@gmail.com*

ABSTRACT

Social media is a platform where people talk and voice their opinion about various problems. To have a perspective as to how many people express their views and observations about diseases on social media, we have various techniques to analyze health care through social media. In this chapter, we use R to perform the analytics on the social media data in order to draw meaningful inferences. The corpus data is collected from popular social media platform Twitter using its API. Locale specific data analytics makes use of geocode of countries like India, Singapore, Indonesia, the UK, and the USA. Tweets of disease-by-disease classes like Diabetes, Stroke, and cancer are analyzed. The resultant analytics are expressed using projections like word count frequency of occurring illustrated as a bar graph.

12.1 INTRODUCTION

Social media refers to online communication through various websites and applications that enable users to create and share content, in order to have a social network. It's a form of electronic communication which enables users to exchange ideas, create social communities, instantly share messages through various applications, etc. Social media has made this globalized world a smaller and a better place to live in as it permits any individual to connect with anyone at any time period all around the globe.

Web-based life refers to transfer of different perspectives by using different existing social platforms like Facebook, Twitter, YouTube, WhatsApp, Instagram, and so on. In another word, we can say that it is a PC intervened innovation that furnishes us with the creating and sharing of thoughts, data, by means of virtual networks and systems.

Web-based social networking also includes transferring photographs etc however; this decreases its genuine utilization.

12.1.1 FEATURES OF SOCIAL MEDIA

Following are the basic features of social media:

1. **User Account:** Creating an account on a social platform comes as a resource for social communication or interaction as without the user account we really can't share the information or communicate with others.
2. **Profile:** This page is one of the necessary activities for identifying the Individual. Basically, it includes the information of the user like profile picture, various posts, recent activity, recommendation, and much more.
3. **Groups, Followers, Friends, Hash-Tags:** The user uses them to make and find connections.
4. **News Feeds:** It refers to the latest posts about the activities going around the globe on the real-time basis.
5. **Personalization:** This permits the user to change their profile photos, personal information like birth date, phone number, etc.
6. **Information Posting or Saving, Updating:** Permits the user to post anything and everything, also the user has an advantage of linking other sites by posting respective URL's on news feeds or only a simple text-based message too works at times.
7. **Comment Section and Like Buttons:** This feature is commonly used to share reaction on a certain post hitting like button or leaving a comment user's post.

12.1.2 TYPES OF SOCIAL MEDIA

Social media can be of different types depending on the content which is being exchanged or distributed. Some types are given as below:

1. **The Social Networking Sites:** There are various social networking sites that help in developing a strong relationship with people. Facebook is one of the leading platforms for socializing. Such social platforms also help the digital marketers to connect with their clients and narrow the bridge between them.

    Messaging and Image sharing sites, in this globalized era where visual content is more beneficial than text. While everyone is leading a fast life, passing on information visually saves time and is easy to understand. Instagram, SnapChat, Pinterest, etc. are popular platforms for online advertisements and social media marketing.

    Videos Sharing Sites: It is much like image-based sites, video hubs like YouTube and Vimeo are attracting the users to put their visual content on it.

    This also helps the education system by sharing the lectures on these sites. Some examples of these sites are YouTube, Vimeo.

2. **Social Blogging:** In the era of social networking, many deprecators may say that traditional blogging is done in the way of the dinosaur. But it's not all true in-spite of various social sites the blogging sites have also make a powerful impact on user's mind. Social sites like Tumblr and Medium have emerged to provide a platform for digital marketers to attract their customers through written content.

3. **Discussion Sites and Social Community:** There are various discussion and social community sites that replace the concept of traditional forums. These types of sites are based on the never-ending process of question and answer. Example of these sites is Yahoo! Answer, Reddit, and Quora.

4. **Mobile Social Media:** There is far more proliferation of social media use by way of applications in mobile devices using the social media via desktop applications. Mobile devices like smartphone and tablet, etc. play a key role in capturing information at live events and events such as micro-blogging using the smartphones have augured the creation, and exchange of user-generated content.

The rich media content generation can assist companies with communication, aid in marketing research, and specific customer base relationship development. This rich-media provides various co-ordinates such as location, network association, and collaboration across various formats such as text, video, audio, etc.

12.2   RELATED WORKS IN HEALTHCARE

Ginsberg et al. (2009) discussed that the seasonal influenza epidemic is a major public health problem that causes tens of millions of respiratory diseases each year and 250,000 to 500,000 deaths worldwide. A way to improve early detection is to monitor the integrity search behavior as queries to online search engines, sent daily by millions of users worldwide. The author presented a method to analyze many Google search queries to track influenza-like illness in the population. This method can use search queries to detect outbreaks of influenza in areas with many users searching the web (Achrekar et al., 2012) discussed in this paper that most previous influenza prediction models were tested using historical influenza status (ILI) data from the Center for disease control and prevention (CDC). Experiments with the data at the time of the forecast showed that models based on data from Twitter are able to reduce forecast errors by a ratio of 17 to 30%, while reference levels only use historical data. The Twitter data is able to produce a forecast two to four weeks before the reference model. In addition, they found that models that use Twitter data on average were more predictive of influenza outbreaks than popular cybernetic data source models of Google's influenza trends (Signorini et al., 2011) investigated the use of information included in the Twitter feed to track the rapid evolution of public opinion on influenza A (H1N1) or swine flu, and to monitor and measure activity disease (Doan et al., 2012) have developed a new method of filtering messages related to the influenza analog (ILI) using 587 million Twitter Weibo messages. Authors first filter the message based on the keyword syndrome in BioCaster Ontology, an existing knowledge model for secular terminology. Then they filter the messages according to semantic features such as negation, subject labels, emoji, humor, and geography.

De Choudhury et al. (2013) quantify postpartum changes in 376 mothers based on dimensions of social engagement, emotions, social networks, and language style. The predictive model classifies mothers who have undergone significant changes after delivery, with a 71% accuracy rate, using observations of prenatal behavior and using 80–83 precision when using postnatal data for 2 to 3 weeks after birth.

Duke et al. (2014) state that promoting evidence-based smoking cessation services via social networking sites can increase the use of smokers. Data used and used by social networks for the tobacco control program (TCP) has not been reported. This study examines the use and levels of activity of TCP in social networks, the coverage of TCP sites and the level of participation in the content of the website. Currently, the coverage of PCT national social networking sites is very low and most technical cooperation programs do not promote the existing interactive potential to stop services or use social networks (Gomide et al., 2011) analyzed how the dengue epidemic is reflected by tweets on the Twitter platform about that topic and how that information can be used for surveillance. Dengue fever is an infectious mosquito-borne disease which is a prime cause of mortality in the tropics and subtropics regions. This analysis allows eliminating content that is not related to dengue surveillance (Hawkins et al., 2016) states that identifying and analyzing the content of publications sent to the hospital can provide new real-time quality measures complementing traditional survey-based methods. The evaluation uses Twitter as a complementary data stream to measure the quality of care seen by patients in US hospitals and it compares the state of mind of patients to established quality measures. Tweets that describe the patient's experience at the hospital cover a wide range of aspects of care that can be identified by automated methods.

Longley et al. (2015) espouse a preliminary empirical assessment of the importance of injecting social network data into small geographic areas that transcends traditional geographic demographic data. They try to empirically critique the advantages and disadvantages of using data in social networks, where spatial and temporal granularity values of activity models are contrasted with lack of information about individual attributes. This paper identifies and assesses the biases inherent in social networks in social science research and to use these biases to evaluate their deployment in research applications (Prochaska et al., 2011) looked at Twitter's social networking account activities to stop smoking. The cross-sectional analysis identified 153 active Twitter accounts from 2007 which related to quitting of smoking and reviewed recent account activity for those accounts again in August 2010. Despite the popularity of establishing a social network to quit smoking, many Twitter accounts are no longer active, and the content of tweets is largely inconsistent with clinical guidelines.

Gohil et al. (2018) in their paper reviewed the different methods for tool production used in the 12 research papers they have included and reviewed at the same time. One of the methods, sentiment method, was trained against 0.45% of the whole data (Salas-Zárate et al., 2017) proposed an aspect-level sentiment analysis (SA) method in their paper which is measured by the words around the aspect obtained through different N-gram methods. They did an experiment to evaluate the effectiveness of their method and found out that N-gram around method is most precise (Xiang et al., 2013) presents an epidemic sentiment monitoring system (ESMOS) that can visualize Twitter user's perception and reaction towards different diseases. They came to this conclusion after developing a two-step sentiment classification workflow which can automatically determine a positive or negative tweet (Wei et al., 2015) addresses the problem of creating awareness among people about epidemics. They kept a track of trends in concern about public health but found out that the task is not only expensive but covers a limited percentage (Geller et al., 2016) discusses how to assimilate openly available health data and making physicians, healthcare staff and patients to understand that by their generic model which assimilates varied open health data sources. They created the model through linked data principles and Semantic Web Technologies (Raghupathi and Raghupathi, 2014) big data (BD) analytics in healthcare is developing into a very exciting field for giving incisive results from exceptionally huge data sets and improving the results while diminishing the expenses. While the potential is incredible; anyway, there are difficulties to counter (Tuli et al., 2018) in their paper provided a description of patient experience sentiments across the United States on Twitter in the last 4 years. Their statistical analyses were done to figure out the differences between patient's experience in metropolitan and non-metropolitan areas (Loeb et al., 2013) discussed the use of social media, especially Twitter, among the American Urological Association (AUA) members as the use has been expanding rapidly. They found out that the use itself is helping the urology trainees and urologist a lot and helps them in urology conferences as well.

12.3 SOCIAL MEDIA DATA PROCESSING

12.3.1 *DATA COLLECTION USING TWITTER*

Data collection is a process of collecting or gathering the information from multiple sources to get the accurate and complete picture of the answers to any problem, evaluate the outcome and test hypothesis.

The following are the steps of collecting data from Twitter.

- **Step-1: Creating a Twitter Application**
  For collecting, the data from the Twitter first make an account on Twitter. If the user already has, an account then goes to the first step i.e., creating Twitter application (Figure 12.1).
  This application basically allows the user to perform analysis by connecting the R console with the Twitter using APIs. Steps for it are as follows:
  Go to https://apps.twitter.com/.

Click on sign-in link and the page will open; enter your log-in id and password (Figure 12.2).

Then the page will appear like shown above (Figure 12.3).

Now click on the Create New App button (Figure 12.4).

Name the application, describe the function of application (Figure 12.5).

12.3.2 *CONNECTING TWITTER APPLICATION WITH R CONSOLE*

For connecting with R, consumer key, consumer secret, access token and access token secrets are the basic requirements. Now go to your R Studio and write some codes for getting the Authentication for searching the tweets.

Set the working directory and install packages like twitteR and ROAuth. Then save the consumer key, consumer secret, access token, and access secret in a variable. Now use the function setup_twitter_oauth() for authentication. When the authentication is done successfully, we can then use the searchTwitter() function of the twitteR library to query the Twitter API to get the tweets. Since the Twitter API has well-defined rate limits, we need to specify a n = 2000, n = 1500, n = 5000 for fetching n number of tweets, else the API will return with an error response.

FIGURE 12.1   Open the Twitter application.

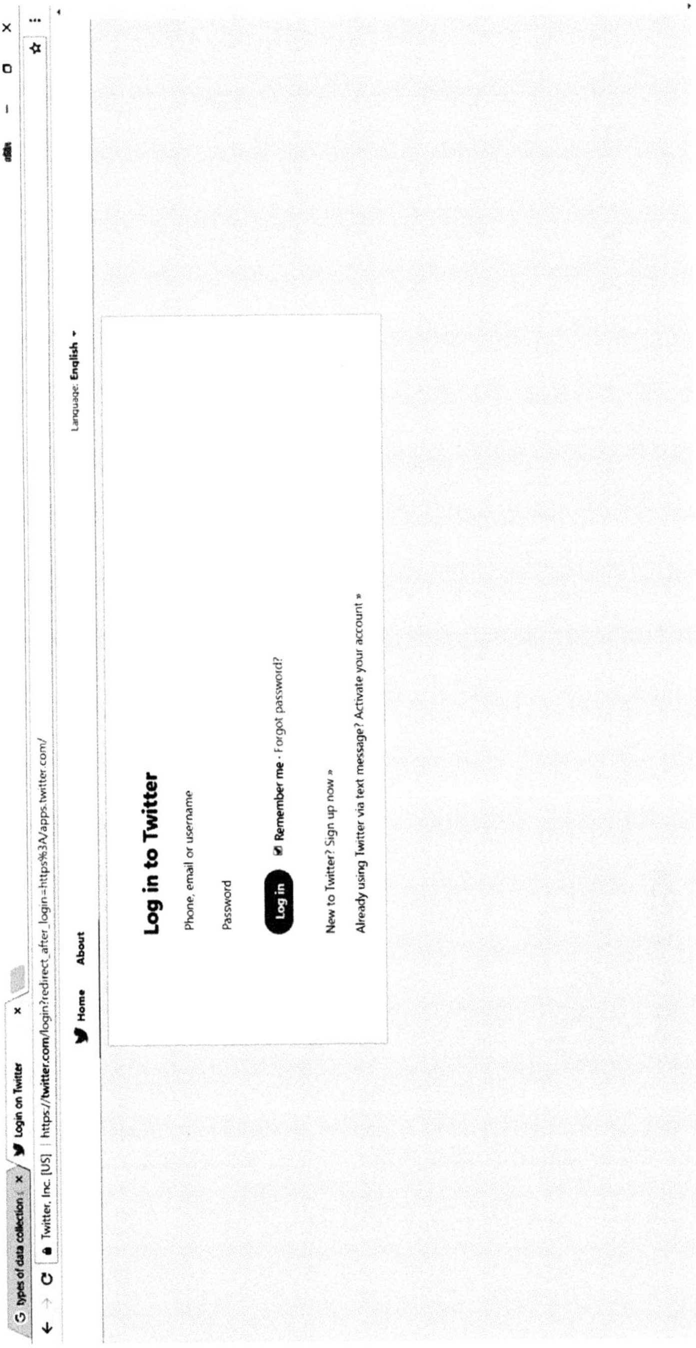

FIGURE 12.2   Log into Twitter.

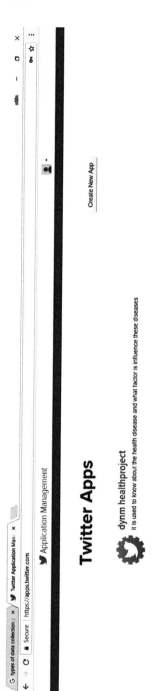

FIGURE 12.3 Twitter page after signing.

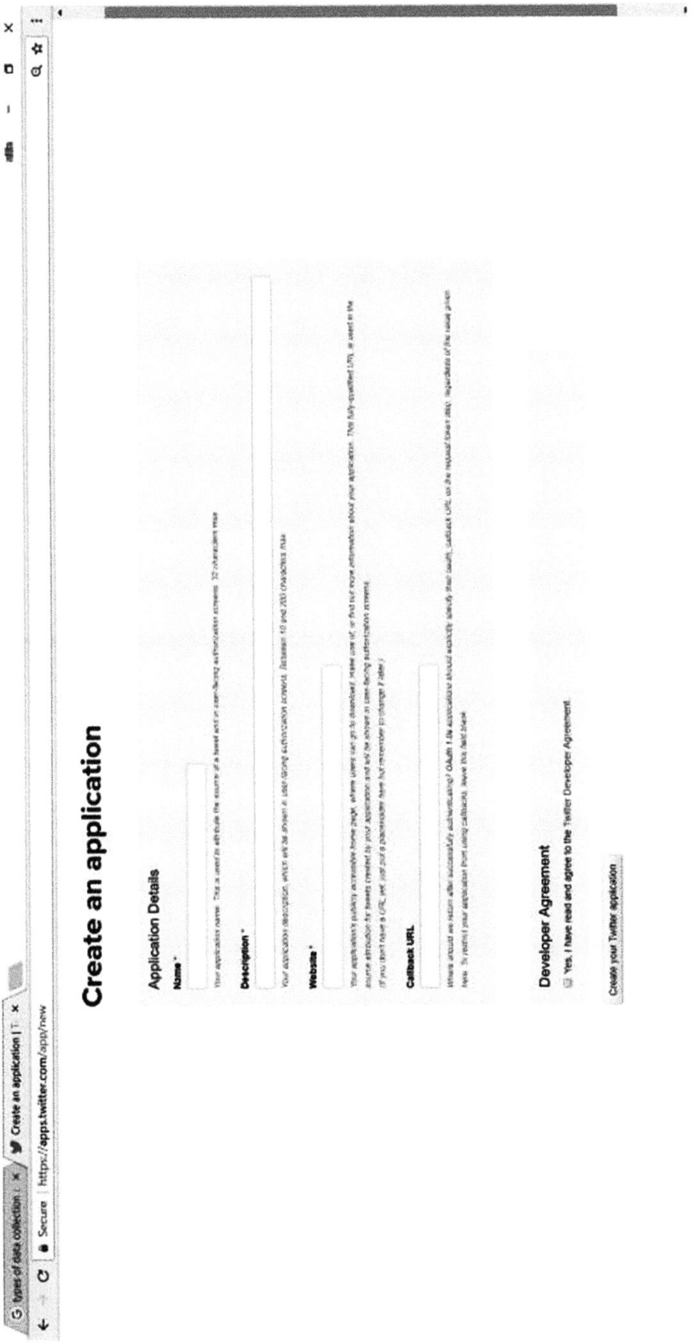

FIGURE 12.4 Creating an application.

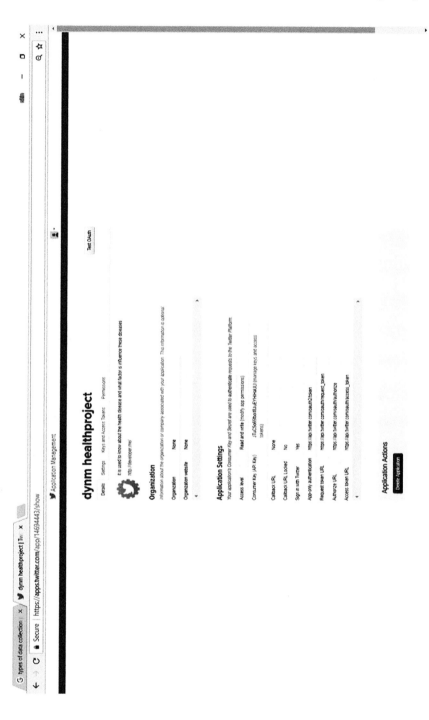

FIGURE 12.5   Page after application creation.

```
library("twitteR")
library("ROAuth")
consumerKey<- 'JTuC5o65fboIf8JuEYHIHaULI'
consumerSecret<-'4GnjtSa1xhvJk5CoG03cOopERNdwPsxx
6bXt'
accessToken<-'278984750-eB1f9Yh2gNHbg0LaDTEDf'
accessSecret<- 'XN2izGYMzhEwQxU5HX588S9ErrLSlrd'
setup_twitter_oauth(consumerKey, consumerSecret, accessToken,
accessSecret)
```

For searching the tweets use the searchTwitter() function it is further explained with the help of code. The example is given below:

```
#extact tweets and create storage file
cancer_list_india<- searchTwitter(cancer_term, n = 2000,geocode
= "20.593684,78.96288,100000mi")
cancer_list_singapore<- searchTwitter(cancer_term, n = 1500,
geocode = "1.352083,103.819836,100000mi")
cancer_df_india<- twListToDF(cancer_list_india)
cancer_df_singapore<- twListToDF(cancer_list_singapore)
write.csv(cancer_stack_india, file = cancer_india, row.names = F)
write.csv(cancer_stack_singapore, file = cancer_singapore, row.
names = F)
```

Above code searches the tweets on cancer with the help of searchTwitter() function. As the Twitter API is quite configurable in terms of providing parameters for search, we can call the searchTwitter()function with number of tweets (n) i.e., 10000 and the also provide a geocode for locale-specific search of tweets. Further, we also provide a specific search term (in our case its various disease or health ailment keywords like Diabetes, Cancer, stroke etc). For converting to data frame twListToDF() function is used which puts the list variable and saves it on the variable tweets.df. After converting the list into data frame save the tweets into the folder by using the write.csv.

12.3.3 DATA CLEANING

It is the process of correcting, removing, or detecting inaccurate or corrupt records from the table, record sets, or datasets. Also, it identifies incomplete, inaccurate, incorrect, or irrelevant parts of data further modifying, replacing the data found as junk. There are various methods to clean the data for making it useful but the simple one uses global substitute function.

For this, first convert the data into the text format by using this code.

```
#use the sapply function to convert into text
cancer.text<- sapply(cancer, function(x) x$getText())
```

Then convert the upper case into a lower case

```
#convert all text to lower case
cancer.text<- tolower(cancer.text)
Then remove all the punctuation, spaces.
#create replacepunctuation function
replacePunctuation<- function(x)
{
x <- tolower(x)
x <- gsub("[.]+[ ],"",x)
x <- gsub("[:]+[ ],"",x)
x <- gsub("[?],"",x)
x <- gsub("[!],"",x)
x <- gsub("[;],"",x)
x <- gsub("[,],"",x)
X <- gsub("rt," ,""x)
X <- gsub("@\\w+," ,""x)
X <- gsub("http\\w+," ,"" x)
x <- gsub("[ |\t]{2,}," ,""x)
x <- gsub("[[:punct:]]," ,"" x)
x
}
```

And invoke the replacePunctuation function on the relevant corpus.

```
cancer_india$text<- replacePunctuation(cancer_india$text)
```

Or the 'tm' library can also be used to remove all the punctuation and it also helps in removing all the stop-words (the, was, were etc.). The Code is given below:

```
cancer.text<- sapply(cancer, function(x) x$getText())
```

Then use the 'tm' library function in R. 'tm' stands for 'term map' In this first convert the text data into corpus then use a tm_map function to remove punctuation, whitespaces, and the stop-words.

```
library("tm")
cancer_india$text<- replacePunctuation(cancer_india$text)
#convert data into corpus
cancer_india.corpus<- Corpus(VectorSource(as.vector(cancer_
india $text)))
#remove stop words
cancer_india.corpus<- tm_map(cancer_india.corpus,
removeWords, stopwords('english'))
```

12.4 DATA ANALYSIS USING R

Reading the whole text and understanding it is a challenging task. Therefore, to make the data easy to understand, graphs are used to conclude the collected data. In R, it is done through graphical representations like bar-graph, pie-chart, etc., or some numeric techniques like regression analysis, hypothesis testing, etc.

12.4.1 PLOTTING BAR GRAPH AND WORD CLOUD

Before the bar graph can be plotted using the barplot() function, the data needs to be rendered as a data frame. The code below illustrates how that is achieved.

1. **Convert the Data into a Data Frame:** In this cleaning processes data is in the form of text, for counting the data and converting the text data into a data frame, the code mentioned below is used. Below is one part of the corpus which has been already tokenized. #convert the corpus to dataframe for wordcount

   ```
   library("SnowballC")
   #india
   cancer_india_dictCorpus<- cancer_india.corpus
   cancer_india.corpus<- tm_map(cancer_india.corpus,
   stemDocument)
   cancer_india_myCorpusTokenized<- lapply(cancer_india.
   corpus, scan_tokenizer)
   # stem complete each token vector
   ```

```
cancer_india_myTokensStemCompleted<-
lapply(cancer_india_myCorpusTokenized, stemCompletion,
cancer_india_dictCorpus)
# concatenate tokens by document, create data frame
cancer_india.dataframe<- data.frame(text = sapply(cancer_
india_myTokensStemCompleted, paste, collapse = " "),
stringsAsFactors = FALSE)
```

For the bar graph, we do a disease-wise breakup of the data in terms of word count.

2. **Counting the Word:** wordcount() function from the ngram library was used for counting the words in the data fetched. The code for this is given below:

```
#counting the data through wordcount
library(ngram)
cancer_india_count<-wordcount(cancer_india.dataframe$text,
sep = " ," count.function = sum)
diabetes_india_count<-wordcount(diabetes_india.
dataframe$text, sep = " ," count.function = sum)
stroke_india_count<-wordcount(stroke_india.dataframe$text,
sep = " ," count.function = sum)
```

Word count frequency Table 12.1 shows high incidence of Diabetes and Cancer in both developing as well as developed countries.

TABLE 12.1   Word Count Frequency for Bar-Graph

Countries	Disease (Range)		
	Diabetes	**Cancer**	**Stroke**
India	26873	7519	7522
Singapore	10108	17176	3448
Indonesia	12673	15303	3045
UK	34030	12197	4363
US	36638	16790	4736

3. **Bar Graph:** barplot() function is used to create the bar graph. All the disease counts are combined and stored in one variable i.e.,

India. In this case, x-axis signifies India's disease parameter and Y axis signifies the disease count (Figure 12.6). The code is given below:

```
#Bar graph of the data.
india<- c(cancer_india_count,diabetes_india_count,
stroke_india_count)
names(india) <- c('cancer,"diabetes,"stroke')
barplot(india, las = 1, space = 2, col = "pink," xlab = "Disease,"
ylab = "Count," main = "India's disease parameter")
```

OUTPUT:

INDIA

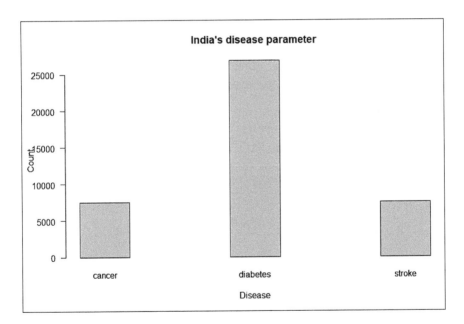

FIGURE 12.6   India's disease parameter.

Similar process is followed for identifying the disease parameters for various other countries (Figures 12.7–12.10).

SINGAPORE

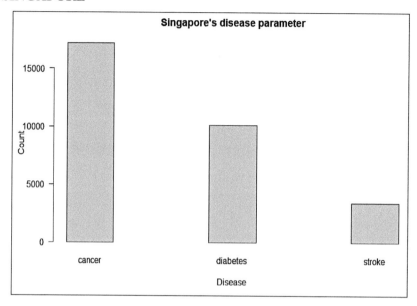

FIGURE 12.7   Singapore's disease parameter.

INDONESIA

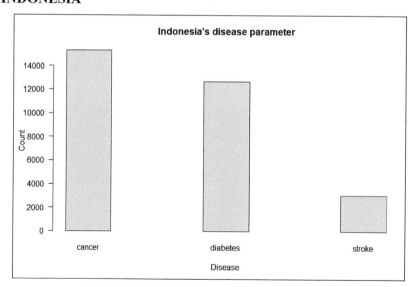

FIGURE 12.8   Indonesia's disease parameter.

UNITED KINGDOM

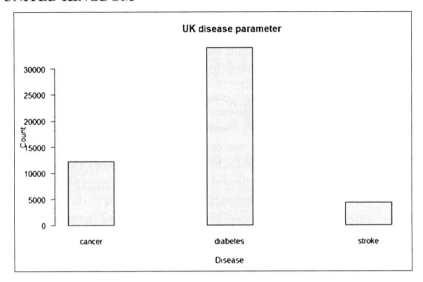

FIGURE 12.9 United Kingdom's disease parameter.

UNITED STATES OF AMERICA

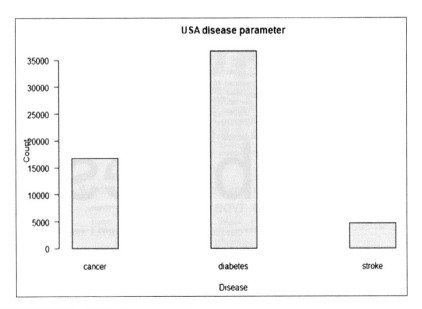

FIGURE 12.10 USA's disease parameter.

12.5   FUTURE PROSPECTS

This chapter presents how social media can be mined for health trends using the R as the mining and statistics tool. This chapter can be enhanced for mining and analyzing social media data at a larger scale and analyzing the health trends by aggregating local location-wise per country. For such analysis, BD technologies like Hadoop would need to be used along with R, with R packages like RHadoop facilitating the same. For faster processing at R side, the fastR package can also be considered for such an analysis.

12.6   CONCLUSION

This chapter looks at how R can be used as an effective social media analysis tool for inferring health analytics such as major prevalent diseases for countries. The data being pulled from streaming data platform such as Twitter and pre-processed and then analyzed in R. R provides all libraries, right from connecting to the social media for streaming data input, to pre-processing, and then to word cloud creation on the data. Further work regarding this is setup of a continuous data input and processing pipeline for near real-time analytics using Kafka as the stream processor and R as the analytics tool. This analysis can be further done for more localized search by geocode and with the use of Kafka, analytics can be done at scale. R also has a large variety of machine learning (ML) libraries which can analyze the input dataset and do predictive analytics.

KEYWORDS

- analytics
- data mining
- R programming
- social media
- tobacco control program
- Twitter

REFERENCES

Achrekar, H., Gandhe, A., Lazarus, R., Yu, S., & Liu, H., (2012). Twitter improves seasonal influenza prediction. In: *International Conference on Health Informatics (HEALTHINF'12)* (pp. 61–70). Vilamoura, Portugal: Nature Publishing Group, based in London, UK.

De Choudhury, M., Scott, C., & Eric, H., (2013). Predicting postpartum changes in emotion and behavior via social media. *Proceedings of the SIGCHI Conference on Human Factors in Computing Systems* (pp. 3267–3276). Paris, France.

Doan S, Ohno-Machado, L., & Collier, N., (2012). Enhancing Twitter data analysis with simple semantic filtering: Example in tracking influenza-like illnesses. *IEEE HISB 2012 Conference*. La Jolla, California, US doi: 10.1109/HISB.2012.21.

Duke, J. C., Hansen, H., Kim, A. E., Curry, L., & Allen, J., (2014). The use of social media by state tobacco control programs to promote smoking cessation: A cross-sectional study. *Journal of Medical Internet Research.*, e169. doi: https://doi.org/10.2196/jmir.3430 (accessed on 16 February 2020).

Ginsberg, J., Mohebbi, M. H., Patel, R. S., Brammer, L., Smolinski, M. S., & Brilliant, L., (2009). Detecting influenza epidemics using search engine query data. *Nature, 457*(7232), 1012–1014. http://dx.doi.org/10.1038/nature0763410.1038/nature07634 (accessed on 16 February 2020).

Gohil, S., Vuik, S., & Darzi, A., (2018). Sentiment analysis of health care tweets: Review of the methods used. *JMIR Public Health and Surveillance, 4*(2).

Gomide, J., Veloso, A., Meira, W., & Almeida, V., (2011). Dengue surveillance based on a computational model of spatio-temporal locality of Twitter. *Proceedings of the 3rd International Web Science Conference* (Vol. 3, No. 1–3, pp. 8). New York, NY, USA.

Hawkins, J. B., et al., (2016). Measuring patient-perceived quality of care in US hospitals using Twitter. *BMJ Quality and Safety. 25*(6), 404–413.

Ji, X., Chun, S. A., & Geller, J., (2013). Monitoring public health concerns using Twitter sentiment classifications. In: *IEEE International Conference on Healthcare Informatics* (pp. 335–344).

Ji, X., Chun, S. A., Cappellari, P., & Geller, J., (2017). Linking and using social media data for enhancing public health analytics. *Journal of Information Science, 43*(2), 221–245.

Ji, X., Chun, S. A., Wei, Z., & Geller, J., (2015). Twitter sentiment classification for measuring public health concerns. *Social Network Analysis and Mining.*

Loeb, S., Bayne, C. E., Frey, C., Davies, B. J., Averch, T. D., & Woo, H. H., (2014). American Urological Association Social Media Work Group. Use of social media in urology: Data from the American Urological Association (AUA). *BJU International, 113*(6), 993–998.

Longley, P. A., Adnan, M., & Lansey, G., (2015). The geotemporal demographics of Twitter usage. *Environment and Planning A., 47*(2), 465–484. doi: https://doi.org/10.1068/a130122p (accessed on 16 February 2020).

Prochaska, J. J., Pechmann, C., Kim, R., & Leonhardt, J. M., (2011). Twitter=quitter? *An Analysis of Twitter Quit Smoking Social Networks. Tobacco Control.* tc.2010.042507. doi: https://doi.org/10.1136/tc.2010.042507 (accessed on 16 February 2020).

Raghupathi, W., & Raghupathi, V., (2014). Big data analytics in healthcare: Promise and potential. *Health Information Science and Systems.*

Salas-Zárate, M. D. P., Medina-Moreira, J., Lagos-Ortiz, K., Luna-Aveiga, H., Rodriguez-Garcia, M. A., & Valencia-Garcia, R., (2017). Sentiment analysis on tweets about diabetes: An aspect-level approach. *Computational and Mathematical Methods in Medicine*.

Sewalk, K. C., Tuli, G., Hswen, Y., Brownstein, J. S., & Hawkins, J. B., (2018). Using Twitter to examine web-based patient experience sentiments in the united states: Longitudinal study. *Journal of Medical Internet Research, 20*(10).

Signorini, A., Segre, A. M., & Polgreen, P. M., (2011). The use of Twitter to track levels of disease activity and public concern in the U.S. during the influenza A H1N1 pandemic. *PLoS One, 6*(5), e19467. doi: 10.1371/journal.pone.

Index